PROBLEM SOLVING IN PHYSIOLOGY

JOEL A. MICHAEL
Rush Presbyterian St. Luke's Medical Center

ALLEN A. ROVICK
Rush Presbyterian St. Luke's Medical Center

PRENTICE HALL, Upper Saddle River, NJ 07458

Executive Editor: *David Brake*
Editorial Project Manager: *Byron D. Smith*
Assistant Vice President, ESM Production and Manufacturing: *David W. Riccardi*
Special Projects Manager: *Barbara A. Murray*
Production Editor: *Mindy De Palma*
Cover Design: *Joe Sengotta*
Production Coordinator: *Ben Smith*
Interior Design and Compositor: *ECL Art*
Cover Photo: *Jim Sullivan, Quill Graphics*

© 1999 by Prentice-Hall, Inc.
Simon & Schuster / A Viacom Company
Upper Saddle River, NJ 07458

Printed in the United States of America

10 9 8 7 6 5 4 3 2

ISBN 0-13-2441047

Prentice-Hall International (UK) Limited, *London*
Prentice-Hall of Australia Pty. Limited, *Sydney*
Prentice-Hall Canada, Inc. *Toronto*
Prentice-Hall Hispanoamericana, S.A., *Mexico*
Prentic-Hall of India Private Limited, *New Dehli*
Prentice-Hall of Japan, Inc.*, Tokyo*
Simon & Schuster Asis Pte. Ltd., *Singapore*
Editora Prentice-Hall do Brasil, Ltda., *Rio de Janeiro*

Contents

Preface

If asked, teachers of physiology will state that their primary educational objective is to help their students "understand" physiology. And there will be near unanimous agreement that "understanding" physiology means acquiring some knowledge and developing the skills to apply that knowledge to accomplish certain tasks (predict the behavior of physiological systems, solve problems, design and carry out experiments, etc.).

Physiology, however, is a difficult subject for most students to understand: physiological systems are made up of many components; complex concepts are required to describe the functions of these components; and there are important interactions between components. Thus, it is also a difficult subject to teach. The number of facts to be learned is enormous and grows larger every day. Authors of textbooks of physiology select those facts that they believe are appropriate to ask a particular student audience to master, and as teachers of physiology, we select the texts that we think best meet the needs of our students. But textbooks are, in general, not an optimal medium to provide students with practice applying these facts to master the process of solving physiology problems.

Problem Solving in Physiology contains a set of physiology problems and answers to be used in assisting students to develop the problem-solving skills that we all believe are an essential part of "understanding" physiology. The problems are a learning resource with which teachers of physiology can engage students in applying their physiology knowledge.

Active Learning and Problem Solving

All too often, reading a textbook or listening to a lecture is a passive process for the student. The primary mental process being engaged is rote memorization. But, if we expect students to organize the facts being memorized so that they can build an understanding of bigger and bigger ideas about the functioning of the body, if we expect students to retain what they memorize, then we must encourage them to make the learning process a more active one. We must provide them with opportunities to consciously build mental models of physiological systems, and we must provide them with opportunities to test the validity of the models that they are building. Accurate mental models give students tools with which to reason about the behavior of physiological systems.

There are many avenues to assisting students to develop active learning skills and many learning environments in which they can practice their developing skills. Traditionally, the laboratory has been one learning environment in which to engage students in an active learning process. Discussion groups, tutorials, and workshops have provided other kinds of environments in which active learning could be stimulated. Solving problems—whatever the physical setting in which this occurs (the laboratory or the small group classroom)—provides students with practice in developing important skills of applying physiology information, helps students integrate the facts of physiology into conceptual models of physiological phenomena, and provides a means for students to test their developing mental models.

Developing a New Approach to Using Problems to Teach Physiology

When we assumed responsibility for teaching physiology to first-year medical students in the 1970s, we immediately began to develop approaches to using physiology problems to foster problem-solving skills and to facilitate student integration of facts. Over the years, we have written problems dealing with essentially all areas of human physiology. Equally important, we have experimented with a variety of procedures for structuring the problems and the problem-solving sessions, with some sessions taking place in the lecture hall with large numbers of students and others occurring in small-group settings. In each of these situations, we have discovered that seemingly small differences in how we do things materially affect the learning that occurs and the student response to their experience. We have shared the problems we have used in our classroom with colleagues at a number of different educational institutions in this country and abroad, and we have discussed how these problems worked (or did not work) with their students. Our ideas about using problem solving in teaching physiology have been presented in a number of national and international forums. The discussions arising from such presentations have helped us further to refine what we do and how we do it. We do not have a universally applicable prescription for how to use problem solving in teaching, but we do know some things that seem to work well in our classroom and in our colleagues' classrooms.

Using This Learning Resource to Achieve Your Objectives

There is a common feature to the format for each of the problems we have written. Each problem is presented in stages. Students are expected to complete each part and discuss the solution to that part before proceeding to the next part of the problem (presented on the next page). Thus every student begins the solution of subsequent parts of the problem with an understanding of the physiology underlying the preceding part. In this way, errors are not allowed to compound, avoiding the danger of students being led astray by an initial failure to understand some aspect of the problem. We have found this organization of our problems to be very important. Compounding failure quickly discourages students from continuing to grapple with the underlying physiology inherent in the problems.

As students discuss parts of the problem and build their mental models, it is important to provide mechanisms with which they can test these models. Two frequently used tools employed in many of our problems are the use of predictions tables and the drawing of concept maps.

The prediction table is a simple way of requesting students to make qualitative predictions (increase, decrease, no change) about the responses of specified components that result from a situation described in the problem. This process requires the students to "exercise" their mental model(s). Then, if their predictions are correct, students gain confidence in the utility of their model(s). If their predictions are incorrect, students can clearly see that their current model(s) need to be changed.

Similarly, constructing a flowchart or concept map of the events involved in a problem requires the students to extract information from their model, and the correctness of their concept map reveals something about the correctness of their models.

The answers we have provided for each problem are intended to be complete in the sense that each step in the solution is developed as explicitly as possible. How one thinks about a physiological issue and the reasoning steps that one must undertake are every bit as important as the answer itself. But, this is *not* a textbook of physiology, and it is not meant to be used like one. When students have difficulties understanding a physiological phenomenon or when questions remain after consulting the answers, it is time to go back to the textbook or the instructor for a more comprehensive discussion of the relevant physiology.

The problems can be used in a variety of different classroom settings. We have used them in small-group problem-solving sessions and in whole-class problem-solving workshops. In both settings, we have worked to ensure certain common features: instructor modeling of problem solutions; instructor facilitation of the students' problem-solving efforts; and active, cooperative problem solving by the students. We have provided descriptions of these classroom formats in the "Owner's Manual" that follows. We do not mean to imply that the approaches we describe are the only way it can done; rather, they are suggestions that can be used and adapted by each instructor. There is no one right way to teach physiology, although there certainly are some broad principles that seem applicable in any course (see Chapters 1 and 2).

Many of the problems are couched in what may seem to be very clinical terms. Yet, these are all problems about physiology, frequently physiology that has been perturbed either by an experimental procedure or by an experiment of nature represented by pathophysiology or pathology in some system. *The problems are not meant to teach medicine and they do not do so*; you do not have to be an M.D. nor does your course have to be part of a medical or allied health curriculum to use these problems in your physiology course! But that such problems are about human physiology or pathophysiology (i.e., they may describe situations in which disease is present) serves to motivate students who are always interested in the processes at work in their own bodies or those of friends and relatives. Where relevant, we have included some information about the clinical context of these as an aid to both the instructor and the student.

Acknowledgments

We want to acknowledge the contributions of current and past colleagues in the Department of Molecular Biophysics and Physiology, specifically Drs. F. Cohen, T. DeCoursey, R. Eisenberg, G. Gottlieb, C. Hegyvary, R. Levis, F. Quandt, E. Rios, and J. Zbilut. Some of the problems were written by other faculty participating in our course. Our faculty colleagues also assisted us in trying out these problems and often suggested revisions that have been incorporated into them. We also appreciate the feedback we have received from the many Rush Medical College students who have used these problems as they attempted to learn physiology.

About the Authors

Joel A. Michael (left) was trained as a neuro-physiologist and conducted research in the neurosciences for twenty years. His involvement in medical education, including thirty years of experience as a physiology teacher, led him to conduct research on the applications of cognitive psychology and computer science to teaching and learning. The worldwide demand for him to conduct faculty development workshops on computer-based education, active learning, and problem solving attests to his expertise in these areas where he has conducted groundbreaking research.

Allen A. Rovick (right) has relied upon forty years of experience as a physiology teacher to develop problem-based laboratories, small-group workshops, and computer-based exercises. His educational research, including studies of tutoring, has culminated in computer teaching programs and a host of other projects, all of which have informed his work on this innovative text.

What Does It Mean To Understand Physiology?

1

Physiology is a discipline taught at a variety of educational levels. In colleges and universities, the courses in which physiology is taught range from introductory or advanced biology to ones that emphasize comparative, insect, general mammalian, or human physiology. Such courses are taught at the most elementary levels to freshmen or sophomores, at more advanced levels to juniors and seniors, and at still more advanced levels to graduate and professional students. These courses may be solely lecture based, may incorporate laboratory work, or may be solely laboratory based.

The instructor for any course is likely to be asked, "What do you expect a student in your course to accomplish?" Students will ask, "What am I expected to learn?" The answers to these questions, whether long and detailed (a list of educational or learning objectives; Bloom, 1958) or brief, is likely to include some statement to the effect that "the student is expected to gain an understanding of physiology."

What exactly does it mean to "understand physiology"? Can one provide some operational or behavioral definition so that a student can tell when he or she has succeeded in meeting this expectation? Equally important, having a definition is essential if it is to be possible to measure whether the students have achieved that goal.

"Understanding physiology" may mean any or all of the following (see Box 1-1) depending on the particular course being taught and the goals of the particular teacher: (1) knowing facts, (2) being able to solve problems, (3) having the skill to carry out certain procedures in the laboratory, (4) being able to design experiments to study physiological phenomena, and analyze the results of experiments to define the components of physiological systems and their behavior, or (5) having the ability to analyze certain problems from a physiological perspective.

Let us attempt to define more precisely what each of these components of understanding involves.

BOX 1-1 "UNDERSTANDING PHYSIOLOGY" MEANS

Possible Components of a Student's Understanding of Physiology

Knowing "facts":
> vocabulary
>
> data
>
> concepts
>
> relationships

Being able to solve problems

Being able to "do" physiology as an experimental science:
> carry out laboratory procedures
>
> design experiments

What Are the Facts of Physiology?

Understanding any discipline requires that students learn the language of that discipline. Thus one particularly important class of facts consists of the specialized *vocabulary* with which experts in the discipline label the objects and behaviors of their study (for example, the resting potential, renal clearance, cardiac contractility). Students must also learn that many words used in everyday discourse may have a special meaning when used to describe physiological phenomenon (force, elasticity). Finally, students need to learn how to use this vocabulary to describe the phenomena encountered in the discipline. Learning the language of physiology involves memorizing the definitions of words but also acquiring the skill to use these words in ways that correctly communicate their intended meaning (whether that meaning is physiologically correct or not).

Another class of facts consists of *data*, the numbers that describe the state of the organism (average, normal blood glucose concentration is 100 mg/100 ml, normal blood pressure is 120/80 mm Hg). *Facts*, isolated pieces of physiology knowledge, must also be acquired (the sinoatrial node is the pacemaker of the heart, digestion of fats requires micelle formation in the small intestine). *Concepts*, larger and more organized pieces of information describing physiological function, are also facts to be acquired (osmotic movement of water, oxygen transport by hemoglobin, the specificity of membrane receptors for hormones). Finally, one must learn the *relationships* between variables (mean arterial pressure = cardiac output $\times$ total peripheral resistance; renal clearance of substance A = urine concentration of A $\times$ urine flow rate/plasma concentration of A).

Although the manner in which students acquire these facts may vary (and one role of the teacher is to provide a number of different ways for students to accomplish this), students are ultimately expected to have these facts available in their memory for retrieval (recall), even if only to answer examination questions, but certainly to use in solving problems.

What Do We Mean by Problem Solving?

Everyone has an intuitive notion of what a "problem" is and what it means to "solve a problem." Nevertheless, it is useful to consider a formal definition of a problem such as the one offered by Smith (1991):

> A problem is a task that requires analysis and reasoning towards a goal (the "solution"); [it] must be based on an understanding of the domain from which the task is drawn; [it] cannot be solved by recall, recognition, reproduction, or applicability of an algorithm alone; and [it] is not determined by how difficult or by how perplexing the task is for the intended solver. (page 14)

Problem solving thus entails having the ability to recognize that you want to "get from here to there," having and being able to access the knowledge required, and having the skills to use that knowledge appropriately to complete the task.

BOX 1-2   PHYSIOLOGY PROBLEMS

Three Different Kinds of Problems Commonly Presented to Students in Physiology Courses

   Quantitative problems
      Calculate the value of something
   Qualitative problems
      Predict the changes that will occur
   Explain
      Describe the causal relationships that lead to some state

---

The problems to be solved in physiology courses are of three kinds (see Box1-2).

1. In *quantitative problems*, the solver must calculate the value of some variable given certain information.

2. Some problems request the solver to make *qualitative predictions* about the direction in which some physiological parameter or parameters will change when a perturbation is introduced in a physiological system.

3. Finally, students are often asked to *explain* some phenomenon, such as to provide a stepwise description of the causal sequence of events that results in an observation to be explained.

We are, of course, distinguishing didactic "problems" or exercises (intended to assist students to learn some part of existing physiological knowledge) from the "problem" that confronts a researcher attempting to gain new knowledge about the functioning of the body or some part of it. The former problems have, arguably, more clearly definable answers (however fuzzy some of the boundaries might be), whereas the latter problems, by definition, do not yet have known answers. We commonly assign problems of the first type to students in courses at all levels, although in advanced undergraduate, graduate, or professional courses we might also pose less structured problems or problems requiring physiological investigation to reach an answer.

Physiology as an Experimental Discipline

Physiology is one of the natural sciences in which it is possible to manipulate nature as a way of uncovering its laws (unlike disciplines such as geology or astronomy, which are largely observational sciences). An instructor might therefore argue that a true understanding of physiology requires that the student gain the skills required to conduct a physiological experiment.

This requirement might only mean that the student is expected to learn how to use common pieces of experimental apparatus (oscilloscopes, amplifiers, strip-chart recorders or polygraphs) or to perform certain procedures (anesthetizing a preparation, preparing it surgically, making up solutions). It can also mean that the student is expected to learn how to design an experiment, starting with the generation of a hypothesis and carrying the process through to the writing up of the results in an acceptable format.

The ability to "do" physiology, to actually carry out experiments, is certainly an important objective of many physiology courses. It is impossible to imagine a student achieving this goal without at the same time learning the vocabulary, facts, and concepts of physiology. It is impossible to imagine that a student could generate and validate a hypothesis without the ability to "solve problems" in physiology. Yet, only a relatively small number of students are expected to "understand" physiology in this way.

Applying a Physiological Analysis to Solving a Problem

An enormous number of problems affect us as individuals and as a society, and many of these problems have a significant physiological component. Such problems are as diverse as personal dietary choices (low fat, low sodium, vitamin supplements) or personal sexual behavior, social issues such as school lunch programs (what is nutritionally sound, what are the consequences of poor nutrition), and environmental pollution. The ability to apply an understanding of the concepts and principles of physiology to problems such as these implies both a fund of knowledge and the skills to use that knowledge to understand still wider phenomena. Further, the ability to solve such problems correctly plays an important role in being able to make personal health decisions as well as contributing to decisions being made by local and national legislatures about health-related matters.

Setting Educational Objectives

We cannot emphasize too strongly that it is the responsibility of each instructor to define the goals for his or her course, goals that include some mix of the above objectives and that may include aspects of problem solving. If students are expected to develop problem-solving skills, it must be understood that these, like other skills, improve with practice. This book will provide a resource with which instructors can help students gain the experience needed to achieve the goals you have set.

References

Bloom, B.S. (editor). 1956. *Taxonomy of educational objectives. Handbook I: Cognitive domain.* New York: David McKay Company, Inc.

Smith, M. U. (1991). A view from biology. In M. U. Smith (Ed.), *Towards a unified theory of problem solving* (pp. 1–19). Hillsdale, NJ: Lawrence Erlbaum Associates.

Solving Physiology Problems | 2

Physiology is a science that seeks to understand the myriad processes that are integrated in a living organism. The first step to acquiring an understanding of some particular physiological phenomenon is the determination of the "entities"—the system parameters and variables—that interact to produce it. As an example, let us consider what happens to a rat whose pancreatic islets (which contain the insulin-secreting β cells) are chemically destroyed with alloxan (a derivative of uric acid). Within a short time the rat will no longer behave normally. It will eat enormous amounts of food but nevertheless lose weight. As this process continues, other visible manifestations of some fundamental abnormality become evident. The rat will be observed to drink large quantities of water and to urinate frequently and copiously. Its breathing will be rapid and deep, that is, it will visibly hyperventilate.

After many years of physiology research, the "entities" involved in the phenomenon described above can now be listed. They include the plasma concentration of hormones such as insulin, glucagon, cortisol and epinephrine as well as the blood concentrations of metabolites such as glucose, free fatty acids, and amino acids. In addition, the rates of uptake (or release) of these substances by tissues such as muscle, liver, and adipose cells are relevant. Box 2-1 lists the "entities" that need to be considered to understand the phenomenon of the alloxan diabetic rat.

BOX 2-1 SOME OF THE "ENTITIES" INVOLVED IN GLUCOSE HOMEOSTASIS

To understand the maintenance of a more or less constant blood glucose level it is necessary to be able to relate the interactions of the entities listed here.

Hormones (plasma concentration):
- insulin
- glucagon
- cortisol
- epinephrine

Metabolites (plasma concentration):
- glucose
- amino acids
- fatty acids
- ketones

Transport rates of metabolites in and out of:
- muscle
- liver
- adipose cells

Furthermore, the qualitative, causal relationships between these "entities" (parameters and variables) can be specified (Figure 2-1) so that the consequences of a change in one parameter on other entities in the system can be predicted. Most textbooks of physiology have such diagrams or their textual equivalent. By understanding some aspect of physiology, we usually mean, in part, that the relationships between entities is known (is in memory) and is usable (can be recalled and applied).

Physiology as a science does not stop when it has determined what parameters and variables interact with each other. Once the causal relationships are understood, the next step is to determine the quantitative relationships between the relevant variables. These relationships may be represented by tables or graphs filled with data points as are quite common in our textbooks. Figure 2-2 is an example of such a graphical representation of the relationship between glucose concentration and stimulated insulin release and the additive effective of gastric inhibitory protein (GIP). Where possible, however, the exact relationships are expressed by mathematical equations relating the relevant variables and necessary physicochemical constants.

To turn to respiratory physiology for an example, the factors that determine the steady-state value of the partial pressure of oxygen in the alveolar space (P_AO_2) are the partial pressure of oxygen in the inspired air (P_IO_2), the alveolar ventilatory rate ($\dot{V}_A$), and the rate at which oxygen is being consumed by the cells of the body ($\dot{V}O_2$). The alveolar gas equation for oxygen expresses these relationships in a most compact form:

$$P_AO_2 = P_IO_2 - \frac{k\dot{V}_2}{\dot{V}_A}$$

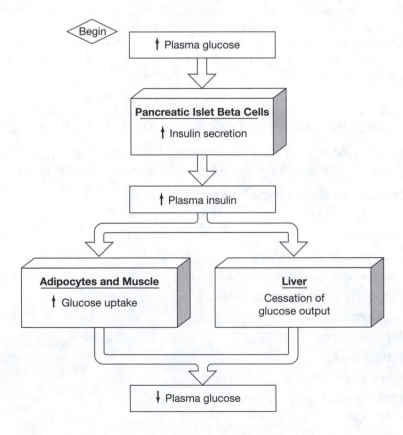

Figure 2-1.   A representation of some of the causal relationships involved in maintaining a normal blood glucose concentration. (From Vander, Sherman, Luciano.)

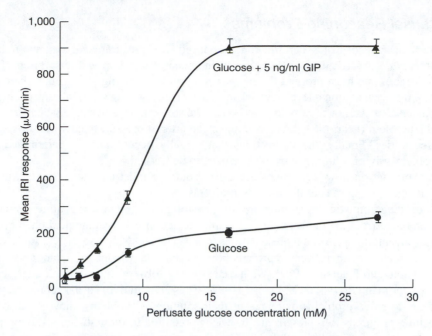

Figure 2-2. The relationship between blood glucose concentration, plasma concentration of gastric inhibitory protein (GIP) and insulin release. (From Patton et al.)

The advantage of having such an equation is that one can reason with it qualitatively (if alveolar ventilation changes, how will P_AO_2 change?) as well as use it to determine quantitative responses of the system to a particular disturbance (if the alveolar ventilation decreases by 25%, what value of P_AO_2 will be present?).

Given this description of physiology as a science, what can we say about the kinds of problems that are presented to be solved, whether as an exercise in problem solving for its own right or as a means of assisting solvers to consolidate their understanding of physiology? As described in Chapter 1, there are three types of problems to be solved, and each requires the solver to be proficient at somewhat different sets of skills to be successful.

1. Problems may involve reasoning about the qualitative, causal relationships between the "entities" making up a particular system. Such problems are frequently posed in terms of *predicting* the qualitative change (increase/decrease/no change) of the components of the system as they respond to some perturbation or disturbance acting on it (Rovick and Michael, 1992). The ability to make such predictions requires both a knowledge of the causal relationships between the entities of the system and the ability to use those relationships to solve the problem. The rules for solving such problems are addressed below.

2. A second kind of problem requires the production of an *explanation* of some physiological phenomenon. What this generally means is that the solver is to construct the sequence of causal steps between relevant "entities" that describes the phenomenon under discussion. Such a sequence of causal relationships can be described at different levels of organization (thus involving different "entities" such as cells, tissues, or organs), and one of the important skills to be acquired by students is identifying the appropriate level to use in making an explanation.

3. Finally, there are *quantitative* problems to be solved. Most of these problems require manipulating an equation, or set of equations substituting the known values, and carrying out the required mathematical operations to arrive at an answer. There are, however, also problems in which information must be obtained from a table or graph and then used to solve the problem. The students may have to first reason through a series of steps to define the relationship(s) to employ before the problem can be solved.

Solving Causal Reasoning Problems

To *predict* the consequences of a perturbation introduced in a physiological system (reduced blood volume) or to *explain* a particular physiological phenomenon (digestion and absorption of fats in the gastrointestinal tract), one must proceed through several problem-solving steps. First, there must be an identification of the organizational level at which an answer is being sought (membrane of sinoatrial node cells, the cardiac pacemaker, or the whole heart). Then the solver must determine which set of entities is involved in the problem (heart rate, stroke volume, mean arterial pressure) and what the relationship is between these entities. Finally, the interactions that actually occur must be determined. Each step involves a problem-solving skill that must be practiced to be learned.

Before elaborating on how these steps are carried out, we must discuss a "tool" whose use can greatly facilitate the solving of causal reasoning problems.

Pictorial or diagrammatic representations of physiological systems are quite common. These illustrations can range from the deliberately cartoonlike depictions of physiological phenomena (Smith, 1991) to more conventional pictorial representations (Mackenna and Callander, 1990; Ackermann, 1992) to highly sophisticated mathematical representations of our knowledge of quantitative physiology (Hopkins, Campbell, and Peterson, 1987). We have evolved a form of diagrammatic representation that we call the concept map (Rovick and Michael, 1992). These qualitative, casual representations can be used by students to solve problems or represent their solution to problems; they can also be used in computer teaching programs to represent the knowledge required to tutor students solving problems at the keyboard. Our approach is akin to that taken by developers of qualitative physics models (see Weld and de Kleer, 1990) rather than that of Novak and Gowin (1984), although our terminology is clearly similar to Novak and Gowin. It is important, however, not to mistake our concept maps for those of Novak and Gowin; ours represent causal relationships, whereas theirs most often represent hierarchical organization of knowledge.

A concept map is made up of three kinds of elements. Physiological parameters are represented by boxes, with the relationship between two parameters represented by an arrow connecting them; the origin of the arrow is on the **determiner** and the head of the arrow is on the **determined** parameter. Thus, in Figure 2-3a, the representation is to be understood to say that X determines the value of Y. It is important to note that such a representation cannot be read in the opposite direction; you cannot say, *given what is represented here*, that Y determines X. (If that were true, it would have to be represented with an arrow in the reverse direction.) The third element needed to describe a relationship between two parameters (Figure 2-3b) is a plus or minus sign indicating whether the relationship is a direct one (if X changes in magnitude, the value of Y will change in the same direction) or an inverse one (if X changes, the value of Y will change in the opposite direction). The plus and minus signs are often placed alongside the arrow head or can be placed in any obvious association with the arrow. (It is not uncommon for the plus sign to be omitted.) A parameter may have any number of determiners, and one parameter may determine any number of others.

A concept map is a useful tool for organizing your understanding of some physiological mechanisms and for making predictions about the behavior of such systems. Consider, for example, the thermoregulatory system that operates to maintain a nearly constant body temperature. This system can be described by the concept map seen in Figure 2-4.

Let us use this concept map to *predict* the changes in body function that will occur during the onset of a fever, when body temperature is increasing. Before reading ahead, predict how cutaneous blood flow, shivering, and sweating will be affected by this disturbance.

Function	Prediction
Cutaneous blood flow	
Shivering	
Sweating	

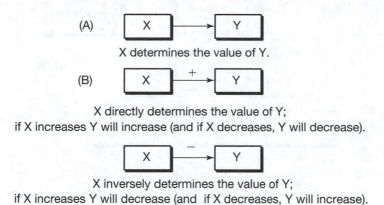

Figure 2-3. How to represent physiological cause-and-effect relationships in a concept map and how to read such a concept map.

A fever results from the presence in the body of substances called pyrogens. These cause the temperature setpoint in the hypothalamus to be reset to a higher value. As a consequence, the thermoregulatory center in the hypothalamus reacts as though the existing core temperature (signaled by the firing rate of the thermoreceptors) is too low. The hypothalamic thermoregulating system then causes changes that will result in the conservation of heat and increased heat production, thus increasing the core temperature. This increase in temperature is accomplished by reducing the rate of sweating and by causing shivering (increased metabolic heat production) and cutaneous vasoconstriction (reducing blood flow to the skin, resulting in less heat loss to the environment) to occur. These changes increase the core temperature; that is, they cause a fever to appear. These responses are illustrated in Figure 2-5.

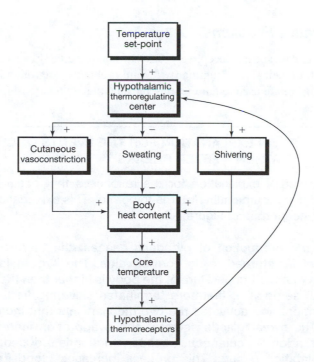

Figure 2-4. A concept map representing the thermoregulatory system.

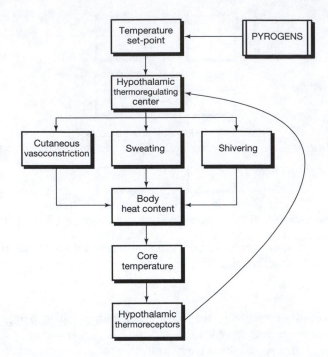

Figure 2-5.   How to predict the responses of the thermoregulatory system to the onset of a fever (the period when body temperature is increasing).

A concept map can be used in another way as well. When a fever breaks and the body temperature returns to normal, the patient will sweat profusely, and a fair-skinned individual will appear quite flushed. If asked to *explain* these "signs," a concept map can be used to represent the responses that give rise to them (Figure 2-6) or to generate a verbal description of these events (Box 2-2).

Solving Quantitative Problems

Quantitative problems can appear deceptively easy to solve, certainly much easier than qualitative problems, due in part, to their greater familiarity. "Word problems" are routinely assigned in arithmetic or science. In addition, we are used to representing the quantitative relationships between variables with

---

BOX 2-2   AN EXPLANATION FOR THE EVENTS OCCURRING WHEN A FEVER BREAKS

Below is a verbal explanation for the responses that occur when a fever breaks and body temperature returns to normal. This explanation can be read from the concept map of Figure 2-6.

When the production of pyrogens ceases, the thermoregulatory setpoint is returned to its normal value. The hypothalamus now senses that the temperature of the body is higher than the (new, but normal) setpoint. It therefore terminates shivering (reducing heat production) and activates the two mechanisms that increase heat loss. The sweat glands secrete, and evaporation increases heat loss. In addition, cutaneous vasoconstriction is reduced; that is, a vasodilatation occurs. This process increases blood flow to the skin, resulting in a greater loss of heat by radiation, conduction, and convection. It also causes the individual to appear flushed.

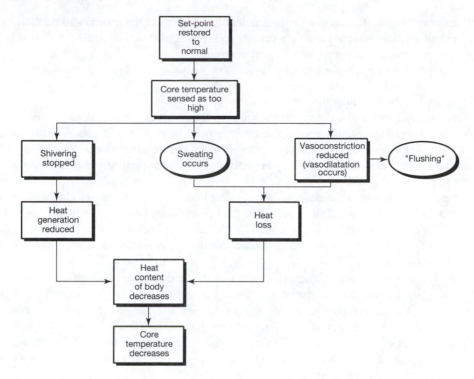

Figure 2-6. A concept map representing the explanation for the responses that occur when a fever breaks (body temperature returning to normal).

equations. Nevertheless, several steps are required to solve such problems successfully, and it will be helpful to review them here (see Box 2-3).

The first step is to read the problem carefully to determine what is to be calculated. It is easy to confuse things that are clearly related and thus calculate a value for the wrong quantity. For example, arterial oxygen content is not the same as arterial oxygen partial pressure; stroke volume and cardiac output are both quantities describing the output of the heart, but they are not the same thing; the renal clearance of a substance is not the same as its rate of excretion. Having identified what to calculate, it is useful to determine the units of measure for that parameter.

BOX 2-3 STEPS IN SOLVING A QUANTITATIVE PROBLEM

1. Determine what you are being asked to calculate (and what units of measure you would expect to arrive at).

2. Analyze the qualitative, causal relationships that determine the value of the quantity to be calculated.

3. Retrieve or derive an appropriate relationship (equation) with which you can make the requested calculation. (Your qualitative analysis may help you remember the equation correctly.)

4. Manipulate the equation algebraically (if need be), and carry out the arithmetic correctly.

5. Check your answer to be sure that your calculations resulted in an answer with the correct units and that the answer makes quantitative sense.

Next, one should think about the parameters that determine the quantity to be determined and how they relate to it. This thought process is another application of causal reasoning, for what is required here is not the retrieval of an equation (that will come in the next step) but the retrieval of the qualitative, causal relationships between interacting parameters. This step often provides a good way to check your conclusion about what is to be calculated because it can make visible those closely related but fundamentally different quantities described above.

Once the quantities that determine the value of the parameter to be calculated have been identified, an equation that quantitatively relates these things must be recalled. Once again, the value of thinking about the problem qualitatively before attempting to find an equation can become evident. One may not remember the equation exactly but may nevertheless be able to reconstruct it knowing what parameters are involved.

Some manipulation of the equation may be required, which may call for algebraic rearrangement of the terms followed by substitution of values and actual calculation, or only the second step may be required. It is very important that you include the units of measure for each of the quantities in the equation. A correct value needs to be generated, but it must also have the correct units for the requested parameter value.

Finally, one should check the answer. First, one should determine whether the units that were derived in the calculation correspond to the units of measure used to describe the parameter asked to calculate; if asked to calculate cardiac output (ml/min or L/min) and the answer is X ml, there is a mistake somewhere. Second, one should determine whether the answer "makes sense." If the Fick equation is being used to calculate the cardiac output of a patient and the answer is 54 ml/min, one can surely conclude that something went wrong in arriving at this answer because it does not make sense (cardiac output is almost always in excess of 3000 ml/min).

Let us apply these problem-solving steps to a specific quantitative problem. The following data describing kidney function in an individual are taken from a problem in this text (Renal 1).

PARAMETER	VALUE
V (urine flow rate)	1.2 ml/min
U_{In} (urine inulin concentration)	0.151 mg/ml
P_{In} (plasma inulin concentration)	15.8 mg/ml
P_{Na} (plasma sodium concentration)	136 mM
U_{Na} (urine sodium concentration)	128 mmoles/L

Question: What percentage of the filtered sodium is reabsorbed by the kidneys?

What follows is the answer to this question as it would be developed in the answer section of this text (in **bold** print as is the case there).

Sodium is freely filtered at the glomerulus, and it is actively reabsorbed by the nephron (returned to the blood).

1. **The rate at which sodium is filtered (the filtered load, FL, in mM/min) is determined by its concentration in the plasma and the rate at which plasma is filtered, the glomerular filtration rate, or GFR:**

$$FL_{Na} \text{ (mM/min)} = P_{Na} \text{ (mM/L)} \times GFR \text{ (ml/min)}$$

2. **The rate at which sodium is being excreted (the excreted load, EL) depends on its concentration in the urine and the urine flow rate:**

$$EL_{Na} \text{ (mM/min)} = U_{Na} \text{ (mM/L)} \times V \text{ (ml/min)}$$

3. The difference between the filtered load (FL) and the excreted load (EL) must represent the rate at which sodium is being reabsorbed:

$$\text{reabsorption rate of Na} = FL_{Na} - EL_{Na}$$

4. The ratio of the rate of reabsorption and the filtered load will then give us the percentage of the filtered sodium that is reabsorbed by the kidney:

$$\% \text{ Na reabsorbed} = \frac{(FL_{Na} - EL_{Na})}{FL_{Na}} \times 100\%$$

Thus, we have completed steps 1 and 2 in solving a quantitative problem (see Box 2-3).

Statements 1 to 4 allow us to write a simple equation relating the quantities above:

$$\% \text{ Na reabsorbed} = 100\% \times \frac{(FL_{Na} - EL_{Na})}{FL_{Na}}$$

To solve this equation, we need to know the value of FL_{Na} ($GFR \times P_{Na}$), which in turn requires that we know GFR. We are not given the value of GFR, but we are provided with data with which to calculate GFR:

$$GFR = C_{In} = \frac{U_{In}}{P_{In}} \times V.$$

We have thus completed step 3 (Box 2-3).

We have been provided with all the data needed to arrive at a quantitative answer to the problem posed. We need only substitute the appropriate values in the two equations we have written and carry out the arithmetic correctly (step 4).

$$GFR = C_{In} = \frac{U_{In}}{P_{In}} \times V$$

$$= \frac{(15.8 \text{ mg/ml})}{(0.151 \text{ mg/ml})} \times 1.2 \text{ ml/min}$$

$$\underline{GFR = 125.6 \text{ ml/min}}$$

$$\% \text{ Na reabsorbed} = 100\% \times \frac{FL_{Na} - EL_{Na}}{FL_{Na}}$$

$$= 100\% \times [(GFR \times P_{Na}) - (U_{Na} \times V)]$$

$$= 100\% \times [(125.6 \text{ ml/min}) \times (136 \text{ mM}) - (128 \text{ mM})(1.2 \text{ml/min})]$$

$$\underline{\% \text{ Na reabsorbed} = 99.1\%}$$

Does this answer "make sense" physiologically? We know that renal blood flow is approximately 20% of the cardiac output, or about 1000 ml/min. We also know that the filtration fraction (the fraction of the blood flow through the kidney that is filtered at the glomerulus) is approximately 0.15. Thus we can estimate that GFR ought to be around 150 ml/min. Our calculated value for GFR is clearly physiologically reasonable. We also know that under normal conditions the reabsorption of Na is almost complete; thus an answer of >90% is reasonable. We have thus completed step 5 and have solved the problem.

References

Ackermann, U. (1992). *Essentials of human physiology*. St. Louis: Mosby Year Book.

Hopkins, R. H., Campbell, K. B., & Peterson, N. S. (1987). Representations of perceived relations among the properties and variables of a complex system. *IEEE Trans. Systems, Man, and Cybernetics, SMC-17*, 52–60.

Mackenna, B. R. & Callander, R. (1990). *Illustrated physiology*, 5th ed. Edinburgh: Churchill Livingstone.

Novak, J. D., & Gowin, D. B. (1984). *Learning how to learn*. Cambridge (UK): Cambridge University Press.

Patton, H. D., Fuchs, A. F., Hille, B., Scher, A. M., & Steiner, R. (Ed). (1989). *Textbook of Physiology*, Vol. 2, 21st ed. Philadelphia: W. B. Saunders Co.

Rovick, A. R., & Michael, J. A. (1992). The prediction table: A tool for assessing students' knowledge. *American Journal of Physiology, 263 (Advances in Physiology Education, 8)*, S33–S36.

Smith, E. K. M. (1991). *Fluids and electrolytes: A conceptual approach*, 2nd ed. New York: Churchill Livingstone.

Weld, D. S., & de Kleer, J. (1990). *Qualitative reasoning about physical systems*. San Mateo, CA: Morgan Kaufman.

UNIT 1

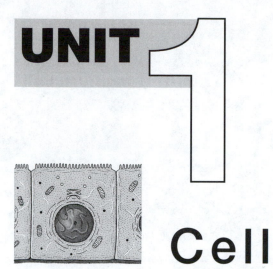

Cell

Problem 1

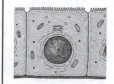

PROBLEM INTRODUCTION

Dehydration is a serious but common condition. It changes body fluid volumes, and sometimes it changes the body fluid composition and concentration. These alterations can cause a redistribution of fluid between compartments which can seriously alter function. This problem deals with the movement of water (osmosis) between red blood cells and the extracellular fluid.

Ed Rivers, a third-year medical student, was alone in the hospital emergency room (ER) one night. It was unusually quiet in the ER, so the residents were getting some much-needed sleep. A patient, Ms. M.B., was brought in showing signs of serious dehydration. Ed tried to give her water but she vomited. Feeling that he must try something else and not wanting to wake the residents, he administered 1 liter (L) of sterile distilled water intravenously (IV).

The questions that follow are aimed at determining the consequences of Ed's action. Assume for simplicity that red blood cells (RBC) contain only solutes to which the RBC membrane is impermeable. The osmolarity of the RBC, the concentration of solute particles, is 300 mOsm/L.

1. Predict the direction of change (increase, decrease, no change) you would expect Ed's infusion to have produced in the listed parameters. Explain your predictions.

PARAMETER	PREDICTION
Ms. M.B.'s plasma osmolarity after the infusion	
Ms. M.B.'s RBC volume after the infusion equilibrates with blood	

 STOP! ANSWER ALL QUESTIONS ABOVE BEFORE PROCEEDING.

2. The volume of Ms. M.B.'s plasma was 3 L before Ed administered the IV. Assume that there was complete mixing of the administered water with her plasma but no mixing with her interstitial fluid. Calculate Ms. M.B.'s plasma osmolarity after the infusion mixed with her plasma but before any water entered her RBC.

3. Assume that Ms. M.B.'s RBC volume was 2 L before the IV was given. Calculate her plasma and RBC osmolarity after the infused water equilibrated between her plasma and her RBC. How much would her RBC volume change? (Assume that there was no mixing with her interstitial fluid.)

STOP! ANSWER ALL QUESTIONS ABOVE BEFORE PROCEEDING.

Ms. M.B. got much worse after Ed's treatment so the resident was called. She drew some blood from the patient and centrifuged it. The denser RBC's were forced to the bottom of the tube. The supernatant plasma was pink. The resident then decided to infuse a sucrose solution into the patient (the RBC membrane is impermeable to sucrose).

4. Why was the patient's plasma pink?

5. Had Ed given Ms. M.B. a sucrose solution instead of water, he might have helped her. What concentration sucrose solution should he have administered to leave her RBC volume unaffected?

STOP! ANSWER ALL QUESTIONS ABOVE BEFORE PROCEEDING.

The resident, knowing what Ed had done, infused the patient with 1 L of a 600 mM (i.e. 600 mOsm/L) sucrose solution. Recall that, after Ed's treatment, the patient's plasma volume was 3.6 L, her RBC volume was 2.4 L, and her osmolarity was 250 mOs/L. Assume that this infusion equilibrated with Ms. M.B.'s plasma and RBCs.

6. Predict (increase, decrease, no change) the effect of the infusion given by the resident on the parameters listed. Explain your predictions.

PARAMETER	PREDICTION
Ms. M.B.'s RBC osmolarity	
Ms. M.B.'s RBC volume	

7. Assuming that none of the administered fluid was excreted, calculate Ms. M.B.'s plasma osmolarity and RBC volume after the resident's infusion had equilibrated with her plasma and RBC.

You have just completed Problem 1 in Unit 1.
Please turn the page to begin work on the next problem.

Problem 2

PROBLEM INTRODUCTION

The intracellular and extracellular concentrations of solutes (electrolytes and nonelectrolytes) have important effects on cell function. Differences in total solute concentration in the two locations affect the distribution of water, whereas ionic distributions and permeabilities affect the resting membrane potential and the action potential.

Figure cell 2-1 shows the distribution of solute inside and outside of a hypothetical cell that is in a steady state. Assume that the cell membrane is freely permeable to water.

Na^+ = 140 mM
K^+ = 5mM
Cl^- = 145 mM
Imperm. non-electrolyte = ?? mM

Na^+ = 20 mM
K^+ = ??mM
Cl^- = 5 mM
A^- (imperm. annion) = 165 mM

OUTSIDE

CELL

Figure cell 2-1. The solute composition of a cell and its extracellular fluid are shown. Question marks (??) represent unknown concentrations.

1. What is the concentration of K^+ inside the cell?

2. If the cell is in a steady state, what concentration of nonelectrolyte must be present outside the cell?

3. Chloride ions are passively distributed across the cell membrane. What is the membrane potential?

4. What is the K^+ equilibrium potential? Is it the same as the membrane potential? If not, why not?

You have just completed Problem 2 in Unit 1.
Please turn the page to begin work on the next problem.

Problem 3

PROBLEM INTRODUCTION

Many axons are covered with myelin, which functions importantly in saltatory nerve conduction. There are, however, conditions in which the myelin sheath around certain axons is destroyed. This problem explores some of the possible consequences of demyelination.

A diagram of an axon with a demyelinated segment is shown in Figure cell 3-1.

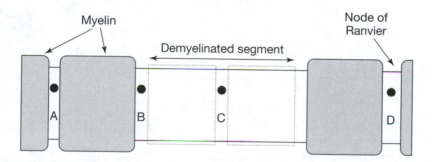

Figure cell 3-1. A segment of an axon. The shaded regions are myelinated. A and D are nodes of Ranvier between myelinated internodes. B is a node of Ranvier lying at the edge of a demyelinated region, and C is a "node of Ranvier" lying completely within the demyelinated region.

Assume that an action potential (AP) is generated at point A and propagates toward point D. Assume also that the channels in the axon membrane of the demyelinated segment are present at the same density and in the same location as in a normal axon. For each of the five parameters listed below, make the qualitative comparisons (greater, less, the same) requested. Explain your predictions. You will find this easiest to do if you first diagram the cause-and-effect relationships that are at work. RP = resting membrane potential, AP = action potential.

PARAMETER	PREDICTION
1. RP at C relative to that at A	
2. Longitudinal current at C when the peak of an AP is at A in the de(myelinated) myelinated axon compared with the current at C in a normally myelinated axon	
3. Transmembrane current at C when an AP is at B compared with the transmembrane current at B when an AP is at A	
4. AP conduction velocity from A to D compared with a normal axon	
5. AP amplitude at D compared with AP amplitude at A	

Explain your predictions here.

1. Resting potential at C relative to A

2. Longitudinal current at C when the peak of an AP is at A in the demyelinated axon compared with the current at C in a normally myelinated axon

3. Transmembrane current at C when an AP is at B compared with the transmembrane current at B when an AP is at A

4. AP conduction velocity from A to D compared with a normal axon

5. AP amplitude at D compared with AP amplitude at A

You have just completed Problem 3 in Unit 1.
Please turn the page to begin work on the next problem.

Problem 4

PROBLEM INTRODUCTION

The transport of many important solutes across epithelial cells is coupled to the transport of Na$^+$, and this process uses the energy content of the Na$^+$ concentration gradient. The intestinal epithelium is one such example.

Suppose that the blood flow to a section of the intestine was severely reduced from normal. Predict the effects (increase, decrease, no change) of this condition on the function of the enterocytes (intestinal mucosal epithelial cells) and on intestinal absorption.

PARAMETER	PREDICTION
1. Activity of the enterocyte Na$^+$/K$^+$-ATPase (Na$^+$ pump)	
2. Lumen to enterocyte Na$^+$ concentration gradient	
3. Glucose absorption from the intestinal lumen	
4. Water absorption from the intestinal lumen	

Explain your predictions here. You will find it helpful to diagram the causal relationships that are involved.

1. Activity of the enterocyte Na$^+$/K$^+$-ATPase (Na$^+$ pump)

2. Lumen to enterocyte Na$^+$ concentration gradient

 STOP! ANSWER ALL QUESTIONS ABOVE BEFORE PROCEEDING.

3. Glucose absorption from the intestinal lumen

4. Water absorption from the intestinal lumen

5. Explain your predictions by diagramming the causal relationships that are involved.

You have just completed Problem 4 in Unit 1.
Please turn the page to begin work on the next problem.

UNIT 2

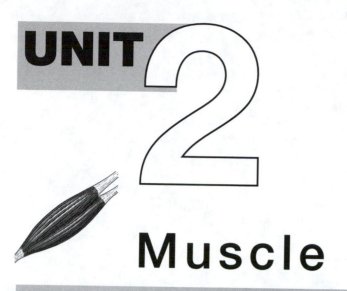

Muscle

Problem 1

PROBLEM INTRODUCTION

The motor pathway consists of a large number of elements organized in a functional sequence. A motor action involves these elements, one after the other, to bring about shortening and force generation by skeletal muscle. Malfunction of any one of these elements can affect muscle strength and coordination.

A patient came to his physician complaining of muscle weakness.

1. List the loci along the complete motor pathway where some defect could result in muscle weakness.

STOP! *ANSWER ALL QUESTIONS ABOVE BEFORE PROCEEDING.*

2. Upon testing the patient, the following observations were made:

   a. Repetitive voluntary activity was associated with normal action potential activity in the motor nerve, although the patient's muscles rapidly fatigue and muscle force declined.

   b. Repetitive direct electrical stimulation of muscle resulted in normally sustained force without evidence of fatigue.

Assuming that a single site of pathology is present, where along the motor pathway is this patient's problem? Explain your reasoning.

3. What might be done to help such patients?

You have just completed Problem 1 in Unit 2.
Please turn the page to begin work on the next problem.

Problem 2

PROBLEM INTRODUCTION

Neural activation of skeletal muscle involves a sequence of steps. Understanding the steps in this sequence helps you to determine the locus of disturbances that affect muscle activation.

You are given an isolated muscle with its motor nerve intact. This preparation is arranged so that you can make electrophysiological measurements from the nerve and the muscle as well as mechanical measurements of the muscle contraction. You observe that if caffeine is added to the fluid bathing the preparation, the peak tension developed in a muscle twitch produced by stimulating the motor nerve is increased.

1. List the *possible* loci at which caffeine may be exerting its effect.

 STOP! ANSWER ALL QUESTIONS ABOVE BEFORE PROCEEDING.

2. Caffeine is known to alter transmembrane movement of Ca^{2+}. What is(are) the possible site(s) at which caffeine is acting?

3. High concentration of caffeine causes an isolated muscle to contract in the absence of any activity in the motor nerve. This contraction is not blocked by nicotinic blocking agents like curare. Where do you think caffeine acts?

You have just completed Problem 2 in Unit 2.
Please turn the page to begin work on the next problem.

Problem 3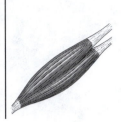

PROBLEM INTRODUCTION

Activation of skeletal muscle occurs through a complex sequence of events, leading to contraction (shortening) or force generation. Abnormal responses at any of these steps will alter muscle function and possibly whole body homeostasis.

Muscle is a biochemical machine that converts chemical bond energy into mechanical force or shortening.

1. Diagram the sequence of steps from the release of transmitter from the motor nerve to the contraction of the skeletal muscle cell that occurs.

2. Biochemical processes give rise to waste products that in many cases pose a homeostatic challenge for the organism. What are the inevitable by-products of muscle contraction with which the body must cope?

 STOP! ANSWER ALL QUESTIONS ABOVE BEFORE PROCEEDING.

A 25-year-old man was brought to the emergency room with a broken leg. His medical history was unremarkable, although he reveals that his mother had died of some "complication" following a hysterectomy.

The patient was taken to surgery for repair of his fractured left distal tibia. He was premedicated with, among other drugs, succinylcholine (a muscle relaxant) and was placed on a ventilator. It was noted that vigorous muscle fasciculation began almost immediately. Nevertheless, general anesthesia was begun with halothane.

3. Succinylcholine binds to the ACh receptors at the motor end plate. Explain the mechanism by which it could produce muscle relaxation.

4. Explain the mechanism by which fasciculations could occur following administration of succinylcholine.

STOP! ANSWER ALL QUESTIONS ABOVE BEFORE PROCEEDING.

There was an immediate tachycardia, and it was noted by the surgeon that the patient had become "as stiff as a board." The anesthetist noted that the patient's core temperature was 39.6° C and seemed to be rising rapidly.

5. What caused the patient to become "stiff as a board"?

6. What could account for the patient's rapid increase in body temperature?

STOP!   ANSWER ALL QUESTIONS ABOVE BEFORE PROCEEDING.

The halothane anesthesia was terminated, and the patient was cooled with ice. Dantrolene was administered intravenously. Oxygen was given.

7. Dantrolene inhibits the release of calcium from the sarcoplasmic reticulum. Explain how this would reverse the patient's hyperthermia.

Laboratory results indicated the presence of respiratory and metabolic acidosis, hyperkalemia, and elevated blood lactate and pyruvate concentrations.

8. Explain these laboratory results.

You have just completed Problem 3 in Unit 2.
Please turn the page to begin work on the next problem.

UNIT 3

Heart Muscle

Problem 1

1. Predict the effects (increase, decrease, no change) of a positive inotropic agent (increases IS) on the following aspects of cardiac muscle contraction.

PARAMETER	PREDICTION
a. rate of isometric force development	
b. rate of isotonic shortening	
c. peak isometric force	
d. peak shortening	
e. duration of contraction	
f. rate of relaxation	

2. Describe the mechanism(s) that cause(s) the inotropic changes in the following contractile properties:

a. rate of isometric force development

b. rate of isotonic shortening

c. peak isometric force

d. peak shortening

e. duration of contraction

f. rate of relaxation

 STOP!   ANSWER ALL QUESTIONS ABOVE BEFORE PROCEEDING.

3. Digitalis (ouabain) is a drug used to counteract the effects of cardiac failure. In heart failure, one or both ventricles are incapable of developing a normal contractile force or of producing a normal stroke volume without an increase in end diastolic filling. In such situations, digitalis increases the inotropic state (contractility) of heart muscle. Digitalis acts by inhibiting the Na^+/K^+ pump.

   a. What effect will treatment with digitalis have on the distribution of ions across the membrane of cardiac muscle cells?

   b. The cardiac muscle membrane contains a secondary active transporter, a Na^+/Ca^{2+} exchanger. This countertransporter is one of the mechanisms that maintains a low cytosolic Ca^{2+} concentration during diastole. It uses the energy content in the transsarcolemmal Na^+ electrochemical gradient to transport 3 Na^+ into the cell for each Ca^{2+} it extrudes from the cell. Explain how the action of digitalis interacts with the action of the transporter to increase the force of contraction of the myocardium.

STOP! ANSWER ALL QUESTIONS ABOVE BEFORE PROCEEDING.

The following studies were carried out on two identical, isolated rabbit hearts. The ventricles were made to contract by electrical stimulation of the ventricular conducting systems. The coronary arteries of heart 1 were perfused with normal, untreated rabbit blood, whereas those of heart 2 were perfused with blood containing a positive inotropic agent (high IS). In the descriptions below, subscript "1" refers to heart 1 and subscript "2" refers to heart 2.

Two experiments were carried out. In the **first experiment**, the ventricles of heart 1 and heart 2 were filled to the same low end diastolic volume (low EDV), and stroke volumes (SV_1 and SV_2) were measured. See the conditions 1 and 3 in the table below.

In the **second experiment** the ventricles of both hearts were filled to an equal but greater EDV (high EDV) than in Experiment 1. Again the stroke volumes were measured (SV_1 and SV_2). See conditions 2 and 4 in the table below.

	EXPERIMENT 1	EXPERIMENT 2
	low EDV	high EDV
normal IS (heart 1)	1	2
high IS (heart 2)	3	4

4. Predict the following (greater, smaller, same):

PARAMETER	PREDICTION
a. the size of SV_1 compared with SV_2 in Experiment 1 (1 versus 3 in the table)	
b. the size of SV_1 with low EDV compared with SV_1 with high EDV (1 versus 2 in the table)	
c. the size of SV_2 with low EDV compared with SV_2 with high EDV (3 versus 4 in the table)	
d. the change (Δ) in SV_1 when going from low EDV to high EDV (Δ 1 to 2 in the table) compared with the change (Δ) in SV_2 when going from low EDV to high EDV (Δ 3 to 4 in the table)	

5. What is the cause of the difference you predicted in Question 4, part a?

STOP! ANSWER ALL QUESTIONS ABOVE BEFORE PROCEEDING.

6. What is the cause of the difference you predicted in question 4, parts b and c?

7. What is the cause of the difference you predicted in question 4, part d?

You have just completed Problem 1 in Unit 3.
Please turn the page to begin work on the next problem.

Problem 2

PROBLEM INTRODUCTION

The contractile performance of heart muscle is determined by its inotropic state, preload and afterload. These factors also determine the stroke output of the heart. The inotropic state is influenced by a variety of factors, including the extracellular Ca^{2+} concentration.

Predict the consequences (increase, decrease, no change) of the acute onset of hypocalcemia on the following.

CARDIAC MUSCLE	PREDICTION
1. velocity of shortening	
2. inotropic state (contractility)	
3. peak force developed	
HEART	**PREDICTION**
4. stroke volume	
5. end systolic volume	
6. end diastolic volume	

CARDIAC MUSCLE

1. velocity of shortening

2. inotropic state (contractility)

3. peak force developed

HEART

4. stroke volume

5. end systolic volume

6. end diastolic volume

You have just completed Problem 2 in Unit 3.
Please turn the page to begin work on the next problem.

Problem 3

PROBLEM INTRODUCTION

The contractile performance of heart muscle depends on its inotropic state, preload, and afterload. These factors also determine the stroke output of the heart. In vivo, the aortic pressure, the pressure against which the left ventricle must pump, is the main afterload. If the outflow resistance of the aortic valve should increase, however, it too can impose a significant afterload on the ventricle.

Predict the consequences (increase, decrease, no change) of an increase in afterload on heart muscle and the effect of a sudden stenosis of the aortic valve (reduction in the cross-sectional area of the valve) on the heart. Explain your predictions.

CARDIAC MUSCLE	PREDICTION
1. velocity of shortening	
2. peak force developed	
3. shortening	
HEART	
4. end systolic volume	
5. end diastolic volume	
6. stroke volume	

CARDIAC MUSCLE

1. velocity of shortening

2. peak force developed

3. shortening

HEART

4. end systolic volume

5. end diastolic volume

6. stroke volume

👉 *You have just completed Problem 3 in Unit 3.*
Please turn the page to begin work on the next problem.

UNIT 4

Nervous System

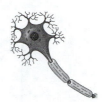

Problem 1

1. Assume that an adequate stimulus of sufficient intensity to elicit action potentials is applied to a purely <u>phasic</u> receptor organ. The receptor organ responds to the rate of change of stimulus intensity. On the same time axis with the stimulus (S) shown in Figure ns 1-1, plot the generator potential (GP) amplitude and the frequency of action potential firing (F).

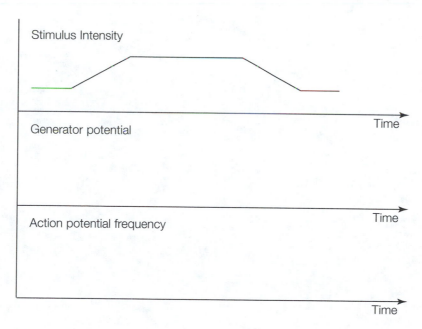

Figure ns 1-1

2. Assume that an adequate stimulus of sufficient intensity to elicit action potentials is applied to a purely <u>tonic</u> receptor organ. The receptor organ response is proportional to the stimulus intensity. On the same time axis with the stimulus (S) shown in Figure ns 1-3, plot the generator potential (GP) amplitude and the frequency of action potential firing (F).

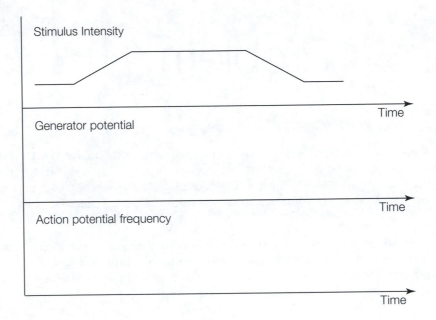

Figure ns 1-3

You have just completed Problem 1 in Unit 4.
Please turn the page to begin work on the next problem.

Problem 2

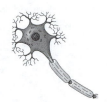

PROBLEM INTRODUCTION

Moment-to-moment regulation of the arterial blood pressure occurs via the baroceptor reflex. The primary receptors for this reflex, the baroceptors, are stretch receptors in the walls of the major arteries. The receptors in the carotid sinus play a primary role in regulating arterial blood pressure in humans.

1. The arterial blood pressure (BP) is the adequate stimulus for the carotid sinus baroceptor. This organ is a mixed receptor whose response has a tonic component that responds in proportion to the absolute pressure and a phasic component that is sensitive to the rate of pressure rise but does not respond to decreases in pressure. Draw the response of the carotid sinus baroceptors on the same time axis as the arterial pressure recording, shown in Figure ns 2-1. Draw the generator potential (GP) amplitude versus time and the individual action potentials (AP) caused by the GP as single vertical lines.

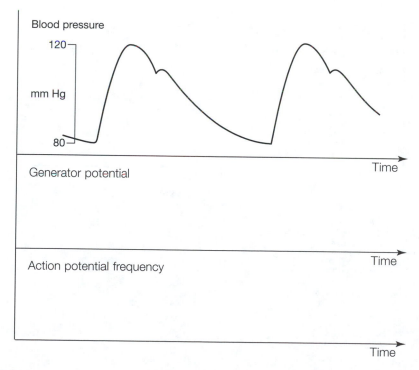

Figure ns 2-1

2. Suppose that the individual whose arterial pressure is shown in Question 1 suffered a coronary occlusion that caused a left ventricular myocardial infarction. As a result, his left ventricular ejection rate and his stroke volume were reduced. Reflex sympathetic activation, however, caused vasoconstriction and increased his heart rate. Therefore, his mean arterial pressure was essentially unchanged from normal see Figure ns 2-2, post-infarct BP. Draw the GP response of his carotid sinus baroceptors and the single APs they cause (as individual vertical lines) after the infarct on the same time axis. Both his pre-infarction BP and GP are shown.

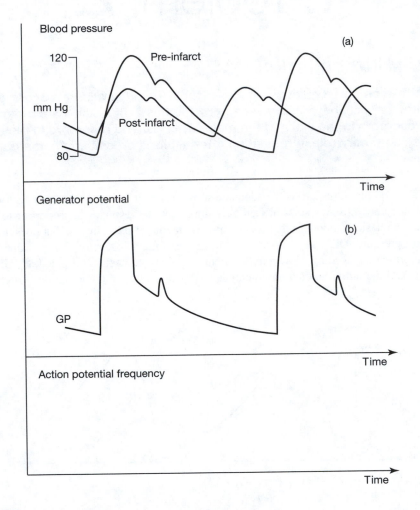

Figure ns 2-2

You have just completed Problem 2 in Unit 4.
Please turn the page to begin work on the next problem.

Problem 3

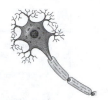

PROBLEM INTRODUCTION

The autonomic nervous system (ANS) is widely distributed to and controls the function of the internal organs. The details of that control have largely been elucidated by use of pharmacological agents that mimic or block the effects of transmitters that are released from the autonomic nerve terminals.

1. An agonist is an agent that reacts with a receptor and mimics the action that normally results from the binding of a transmitter to that receptor. Thus a muscarinic agonist mimics the effects of acetylcholine to a muscarinic receptor. An antagonist blocks the action that normally results from the binding of a transmitter to a particular receptor. Thus a β-adrenergic antagonist blocks the actions of epinephrine (a β-adrenergic agonist) that normally result from its binding to a β-adrenergic receptor.

    Assuming that you had an agonist and an antagonist for every autonomic transmitter receptor, how could you determine which receptor types exist in any autonomically controlled effector?

 STOP! ANSWER ALL QUESTIONS ABOVE BEFORE PROCEEDING.

2. Using the method you defined in Question 1 and your knowledge of the autonomic control of the function of the internal organs, predict the effects (increase, decrease, no change) of the following autonomic agonists on heart rate (HR).

AGONIST	HR CHANGE
α-adrenergic	
β-adrenergic	
muscarinic	
nicotinic	

3. Using autonomic pharmacological agents, how could you determine which receptors in an organ are subjected to a tonic release of neurotransmitter?

STOP!   ANSWER ALL QUESTIONS ABOVE BEFORE PROCEEDING.

4. Using your answer to Question 3 and your knowledge of the autonomic control of heart rate (HR), predict the heart rate changes (increase, decrease, no change) that the pharmacological agents listed in the table below would produce in a resting individual.

ANTAGONIST	HR CHANGE
α-adrenergic	
β-adrenergic	
muscarinic	
nicotinic	

5. Predict the heart rate changes that the agents would produce in a vigorously exercising individual.

ANTAGONIST	HR CHANGE
α-adrenergic	
β-adrenergic	
muscarinic	
nicotinic	

You have just completed Problem 3 in Unit 4.
Please turn the page to begin work on the next problem.

Problem 4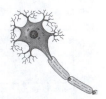

PROBLEM INTRODUCTION

The values of many physiological parameters (plasma concentrations, fluid volumes, blood pressure, etc.) are held within very close limits in the body. Maintenance of this constancy is essential to life and is called homeostasis. Homeostasis involves the nervous and endocrine systems, which act as the integrators and the controllers of the processes that bring about homeostasis.

Homeostatic systems all have a specific set of functional components that interact in a limited number of ways to enable each system to carry out its activity. This problem reviews the types of components that must be present and their interactions.

1. Suppose that the plasma concentration of a substance, X, was regulated (i.e., its value was held within some narrow limits). List the components of a control system that would have to be present to carry out this homeostatic activity and the function that those components must perform.

 STOP! ANSWER ALL QUESTIONS ABOVE BEFORE PROCEEDING.

2. Diagram the interaction(s) between the components in your list, showing how these interactions maintain the plasma concentration of X. Add an external disturbance—for example, an increased loss of X from the body in the urine—and show how the system would minimize the effect of this disturbance. In your diagram, use the convention shown in Figure ns 4-1.

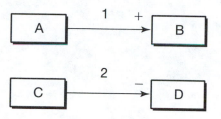

Figure ns 4-1. Panel 1 shows the interaction between two elements in a physiological control system. This representation should be read, "B changes in the same direction as A" or "A acts on B in a direct causal manner." The representation in panel 2 should be read, "D changes in the opposite direction from C" or "C has an inverse causal effect on D." Often, the plus signs are omitted and only the minus signs are shown.

3. What would be the effect on $[X]_{plasma}$ of blocking the normal interaction (connection) between any two of the components in your system? For example, first eliminate the connection between the controller and between the source and the sink.

STOP! *ANSWER ALL QUESTIONS ABOVE BEFORE PROCEEDING.*

4. Blood volume is homeostatically regulated by a neuroendocrine mechanism. Invent a neuroendocrine control system that could carry out this regulatory function. Diagram your invention. You need not specify the detailed mechanism by which specific physiological effects are carried out. You need only state that they occur; for example, A acts on organ B to cause effect C. Try out your control system using some conditions you know to affect blood volume.

Problem 5

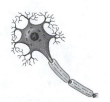

PROBLEM INTRODUCTION

Throughout the nervous system, the predominant mode of neuron-to-neuron and neuron-to-effector (muscle or gland) transmission involves the conversion of an electrical event in the presynaptic neuron (an action potential) to a chemical event (secretion of neurotransmitter) and then back to an electrical event (an action potential) in the postsynaptic cell. Interference with any step in this sequence has significant effects on nerve and/ or effector function.

Botulinus toxin causes paralysis. It binds to presynaptic sites at the motor end plate (neuromuscular junction) and prevents the exocytotic release of transmitter from vesicles. Predict the consequences (increase, decrease, no change) of blocking the release of acetylcholine at the neuromuscular junction on the following.

PARAMETER	PREDICTION
1. Presynaptic resting potential	
2. Presynaptic action potential amplitude	
3. Presynaptic Ca^{2+} influx	
4. Presynaptic transmitter release	
5. Postsynaptic resting potential	
6. Postsynaptic end plate potential	
7. Postsynaptic action potential amplitude	
8. Postsynaptic action potential frequency	
9. Muscle contractile strength	

1. Presynaptic resting potential

2. Presynaptic action potential amplitude

3. Presynaptic Ca^{2+} influx

4. Presynaptic transmitter release

5. Postsynaptic resting potential

6. Postsynaptic end plate potential

7. Postsynaptic action potential amplitude

8. Postsynaptic action potential frequency

9. Muscle contractile force

10. Draw a concept map that shows the steps that occur between a presynaptic action potential and the production of a postsynaptic action potential. Show where in this process botulinus toxin exerts its effect.

 You have just completed Problem 5 in Unit 4.
Please turn the page to begin work on the next problem.

UNIT 5

Cardiovascular

Problem 1

PROBLEM INTRODUCTION

The flow of blood in the circulation is described by the laws of hemodynamics. These laws express the relationship between pressure, flow, and resistance. An understanding of the functions of the heart and circulation requires an understanding of these relationships.

Figure cv 1-1 is a model of the cardiovascular system. It consists of a heart and a circulation that perfuses five organs or tissues (1 through 5) as shown. This system *has no reflexes* and the circulation is assumed to be noncompliant, but in all other ways it behaves like a normal, human cardiovascular (CV) system. The pressure at the aorta (MAP) is maintained constant at 100 mm Hg. Blood flow (Q) through each of the organs or tissues is shown, as are their distances (in centimeters) from the heart.

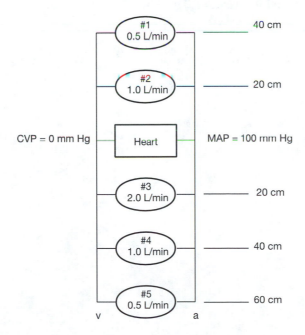

Figure cv 1-1. A model of a simple cardiovascular system with a heart and five organs or tissues (1 through 5) that are perfused by the heart. This system has no reflexes. The blood flow through each organ is indicated, and the distance of the organ from the heart is shown on the right side.

1. What is the resistance R to flow through organ 3?

2. What is the total peripheral resistance (TPR)?

STOP!   *ANSWER ALL QUESTIONS ABOVE BEFORE PROCEEDING.*

The pressure anywhere in the circulation, P, is equal to $P_{pump} + P_{hydrostatic}$, where $P_{hydrostatic} = \rho \times g \times h$. (where ρ is the density of the fluid, g the gravitational constant, and h the height of the column of fluid; $\rho \times g \times h$ for a 1 cm column of blood = 0.75 mmHg).

3. What is the pressure at point a while the individual is lying down? At point v? What is the ΔP across the system $(P_a - P_v)$?

4. With the individual standing quietly, what is the pressure at point a? At point v? What is the ΔP across the system?

 STOP! ANSWER ALL QUESTIONS ABOVE BEFORE PROCEEDING.

Figure cv 1-2 represents the anatomical elements making up one of the organ circulations in the previous model; the relative resistance to flow posed by each vascular component is indicated.

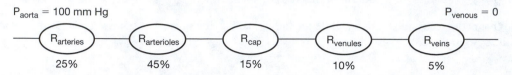

P_{aorta} = 100 mm Hg P_{venous} = 0

| $R_{arteries}$ | $R_{arterioles}$ | R_{cap} | $R_{venules}$ | R_{veins} |
| 25% | 45% | 15% | 10% | 5% |

Figure cv 1-2. A model of the flow path of a single organ in the circulation starting with the aorta and proceeding through all the blood vessels in series to the vena cava. The percentage of the total flow resistance of this path represented by each element is indicated.

5. What is the pressure at the distal end of the arterioles?

6. What will happen to the pressure at the midpoint of the capillary (P_{mc}) if P_{aorta} is increased by 10 mm Hg? If P_{venous} is increased by 10 mmHg?

STOP! ANSWER ALL QUESTIONS ABOVE BEFORE PROCEEDING.

7. Calculate the compliance for each type of vessel at normal P_{artery} and normal P_{vein}.

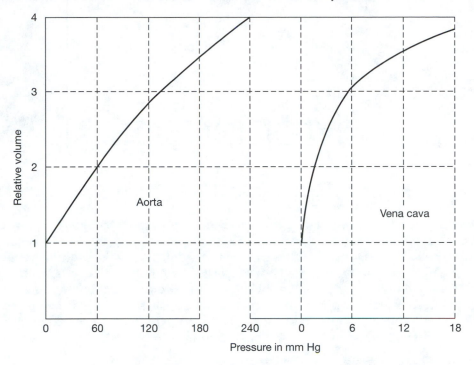

Figure cv 1-3. Pressure-volume (compliance) curves for the aorta and the vena cava.

When the volume contained in an elastic structure undergoes a change (ΔV), the pressure within it changes (ΔP). Conversely, if the pressure within is changed, there will be a corresponding change in volume. The relationship between the change in pressure and the change in volume for an elastic structure is called the <u>compliance</u> (C), which is defined as $\Delta V/\Delta P$.

Figure cv 1-3 illustrates the pressure-volume relationships for the vena cava and the aorta.

You have just completed Problem 1 in Unit 5.
Please turn the page to begin work on the next problem.

Problem 2

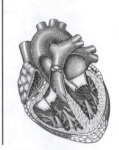

PROBLEM INTRODUCTION

Understanding the integrated functioning of the cardiovascular system—the interrelationships between the heart, the circulation, and the cardiovascular control mechanisms—can be aided by determining how this system behaves when a single part of it functions abnormally. For example, if the cross-sectional area of a valve is reduced, the heart will function differently as a pump, there will be a change in arterial pressure, and this will cause a response of the blood pressure regulating system (the baroreceptor reflex). This reflex response will further alter the function of both the heart and the circulation.

Cardiac Catheterization Procedure

Although some information about the function of such an abnormal cardiovascular system can be obtained "from the outside" (systolic and diastolic pressures with a sphygmomanometer, heart sounds with a stethoscope), much of the information needed to characterize the system is available only from within it. Such information has been gained from experimental studies on animal preparations in which lesions to the cardiovascular system have been produced. Now, however, information about cardiovascular abnormalities is also routinely available from medical diagnostic procedures carried out on human patients where nature has performed the "experiment."

Cardiac catheterization is a common procedure used to diagnose cardiac disease. This invasive procedure is used only when serious disease is suspected, but the risks associated with this procedure have in recent years become minimal. A cardiologist obtains information about the state of a patient's heart by introducing a catheter (a thin plastic tube) into the circulation and from there into the heart. The patient is awake but sedated; the electrocardiogram is monitored continuously (yielding the heart rate), as is the rate of oxygen consumption. The catheter is inserted into a vein, usually the femoral vein in the groin, and is slowly advanced into the right atrium, the right ventricle, the pulmonary artery, then finally "wedged" in a small pulmonary artery. A second catheter is threaded through the femoral artery and aorta into the left ventricle. It may be passed still further into the left atrium.

As the catheters are advanced, blood samples are taken and analyzed for their oxygen content, and pressure measurements are made at different locations within the circulation and the heart. A radio-opaque dye may be injected to permit X-ray visualization of the silhouette of the lumina of the blood vessels or the chambers of the heart. Figure cv 2-1 schematically illustrates the locations from which measurements are usually obtained.

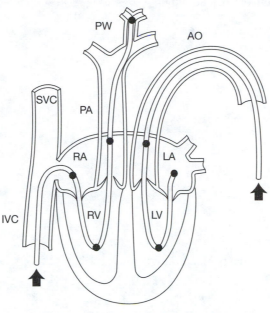

● = Pressure measurement location

Figure cv 2-1. Schematic diagram of the heart and the major vessels. IVC = inferior vena cava; SVC = superior vena cava; RA = right atrium; RV = right ventricle; PA = pulmonary artery; PW = point in pulmonary circulation from which "wedge" pressure is obtained; LA = left atrium; LV = left ventricle; AO = aorta.

> **In this problem, you will examine the consequences of a single specific cardiac abnormality on the function of the total cardiovascular system using information obtained from cardiac catheterization of a patient.**

Mr. A.B., a 56-year-old patient, was admitted to the hospital complaining that he had recently begun to tire on exertion (a round of golf) and that he had occasional heart palpitations. He did not appear to be in acute distress at the time of examination. Physical examination revealed a moderate systolic ejection murmur.

Mr. A.B. was referred for cardiac catheterization, and the following data were obtained:

MEASUREMENT	VALUE
heart rate	76/min
ventilation	5.16 L/min
O_2 consumption	219 ml/min
hemoglobin	13.6 gm/100 ml blood (gm %)
O_2 content: mixed venous blood	13.1 ml O2/100 ml blood (vol %)
O_2 content: pulmonary artery	13.1 vol %
O_2 content: pulmonary vein	17.4 vol %
O_2 content: aorta	17.4 vol %

1. Calculate Mr. A.B.'s left ventricular output.

2. Calculate Mr. A.B.'s pulmonary blood flow.

3. What is Mr. A.B.'s stroke volume?

 STOP!   ANSWER ALL QUESTIONS ABOVE BEFORE PROCEEDING.

Mr. A.B. was diagnosed as having a stenosed aortic valve.

4. Given this pathology, predict the changes in pressure (increase/decrease/no change) that would be measured:

PREDICT PRESSURE CHANGES IN:	INCREASE/DECREASE/NO CHANGE
right atrium (mean)	
right ventricle (systolic)	
left atrium (mean)	
left ventricle (systolic)	
aorta (systolic)	

5. Diagram the causal relationships that account for the patient's systolic ejection murmur.

6. Explain the patient's "normality" at rest and his tiring upon exertion.

You have just completed Problem 2 in Unit 5.
Please turn the page to begin work on the next problem.

Problem 3

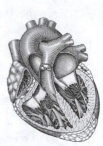

PROBLEM INTRODUCTION

Blood flow through the normal heart is unidirectional, primarily because of the one-way valves in the system. Failure of the valves to function normally or the presence of abnormal flow paths in the heart or its associated vessels alters this normal pattern of flow and results in serious disturbances to normal cardiovascular function. In this problem, we consider an individual with such pathology.

See Cardiovascular 2 for a description of the cardiac catheterization procedure with which the information about this patient was acquired.

Mrs. J.Y., age 35, was admitted to Presbyterian-St. Luke's Hospital to determine the cause of her complaint of vague chest pains in stressful situations. Angina had been ruled out at another hospital, where she had also had had an echocardiogram because of the presence of a heart murmur. The echocardiogram suggested the presence of right ventricular enlargement. She was sent to this particular hospital for further evaluation.

Right ventricular enlargement was confirmed, and examination revealed a moderate systolic murmur and a split second heart sound.

Mrs. J.Y. underwent cardiac catheterization, and the following data were obtained:

MEASUREMENT	VALUE
heart rate	85/min
ventilation	5.46 L/min
O_2 consumption	219 cc/min
hemoglobin	13.4 gm/dl
O_2 content: mixed venous blood	13.14 ml/dl
O_2 content: pulmonary artery	16.31 ml/dl
O_2 content: pulmonary vein	17.78 ml/dl
O_2 content: aorta	17.78 ml/dl

1. Calculate Mrs. J.Y.'s pulmonary blood flow.

2. Calculate Mrs. J.Y.'s systemic blood flow.

3. Compare the two results and explain how they are possible.

STOP! ANSWER ALL QUESTIONS ABOVE BEFORE PROCEEDING.

The following pressures were recorded during cardiac catheterization:

MEASUREMENT	VALUE
right atrium (mean)	6 mm Hg
right ventricle (systolic/diastolic/end diastolic)	30/2/8 mm Hg
pulmonary artery (systolic/diastolic/mean)	30/6/14 mm Hg
left atrium (mean)	6 mm Hg
left ventricle (systolic/diastolic/end diastolic)	106/-3/10 mm Hg
aorta (systolic/diastolic/mean)	102/66/84 mm Hg

4. From the data above determine the location of the abnormal flow path that is present in this patient.

STOP!   ANSWER ALL QUESTIONS ABOVE BEFORE PROCEEDING.

Mrs. J.Y. has an atrial septal defect, a hole in the septum separating the two atria.

5. Given this pathology, explain the following signs and symptoms:
vague chest pains in stressful situations

systolic murmur

split second heart sound

right ventricular enlargement

"normal" MAP

☞ **You have just completed Problem 3 in Unit 5.**
Please turn the page to begin work on the next problem.

Problem 4

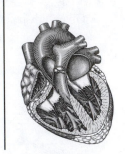

PROBLEM INTRODUCTION

Alterations in function of the cardiovascular system due to an abnormality present in that system frequently give rise to abnormal function in other organ systems. The patient described in this problem presents with just such a mix of signs and symptoms.

See problem Cardiovascular 2 for a description of the cardiac catheterization procedure with which the information about this patient was acquired.

Mrs. V.H. is a 45-year-old, overweight woman who has complained of increasing shortness of breath over the past two or three weeks. She has "three pillow" orthopnea and mild pedal edema. On the day of her hospital admission, she had a cough productive of whitish sputum and a sharp retrosternal pain. On examination, she was found to have diffuse wheezing, associated with acute respiratory distress. Her blood pressure was 180/70 mm Hg, her pulse was 90/min, and her respiratory rate was 26/min. She had a moderate systolic murmur and a diastolic rumble. Her chest X ray showed an enlarged left atrium, pulmonary vascular redistribution (increased filling of her pulmonary blood vessels), and interstitial pulmonary edema. Her ECG showed sinus tachycardia and a suggestion of right ventricular hypertrophy.

Cardiac catheterization was performed, and the following results were obtained:

MEASUREMENT	VALUE
heart rate	82/min
ventilation	11.28 L/min
O_2 consumption	273 cc/min
hemoglobin	14 gm %
(a-v) O_2 difference, pulmonary	7.2 vol %
(a-v) O_2 difference, systemic	7.2 vol %
cardiac output	3.8 L/min
cardiac index	1.8 L/min/M^2
body surface area	2.1 M^2

Pressures measured were:

MEASUREMENT	VALUE
right atrium (mean)	4 mm Hg
right ventricle (systolic/diastolic)	45/5 mm Hg
pulmonary artery (systolic/diastolic/mean)	45/24/31 mm Hg
pulmonary wedge (mean)	18 mm Hg
left ventricle (systolic/diastolic)	110/5 mm Hg
aorta (systolic/diastolic/mean)	109/70/83 mm Hg

1. Calculate the pulmonary vascular resistance (PVR), systemic vascular resistance (SVR), the SVR/PVR ratio.

2. Which of Mrs. V.H.'s cardiovascular values are normal? Which are abnormal?

3. Propose possible abnormalities to account for these data.

✋ ***STOP!   ANSWER ALL QUESTIONS ABOVE BEFORE PROCEEDING.***

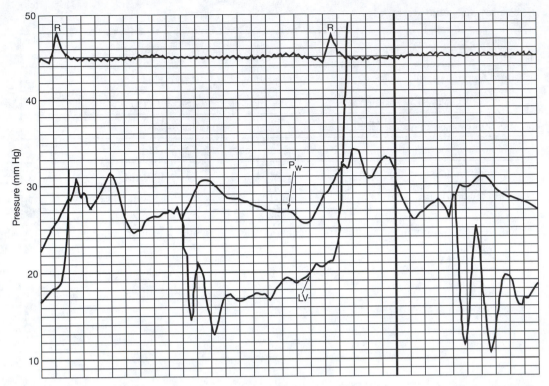

Figure cv 4-1. Tracings obtained during the catheterization of patient Mrs.V.H. The top tracing is the electrocardiogram (ECG). The bottom two tracings are the pulmonary wedge (PW) pressure) and the diastolic component of the left ventricular (LV) pressure (at this scale systolic pressure is "off the top" of the page).

4. Is any aspect of the above recording (Figure cv 4-1) abnormal?

5. Use the information given here to explain the other abnormalities.

6. What is the cause of Mrs. V.H.'s problem?

7. Construct a causal diagram to represent the physiology of Mrs.V. H.'s cardiovascular system.

You have just completed Problem 4 in Unit 5.
Please turn the page to begin work on the next problem.

Problem 5

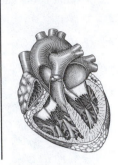

PROBLEM INTRODUCTION

As the posture of the body changes—for example, in going from lying down to standing up—the effects of gravity on cardiovascular function change significantly. Nevertheless, mean arterial pressure is maintained more or less constantly. In this problem, we consider this phenomenon and the mechanisms that generate it.[1]

1. Consider a normal individual who quickly goes from a supine posture to an erect one. Predict the changes (increase/decrease/no change) that will occur in the following cardiovascular parameters. Compare the values of these parameters while erect with the values of the parameters while lying down.

PARAMETER	CHANGE WHILE ERECT
stroke volume	
heart rate	
cardiac output	
total peripheral resistance	
pressure in root of aorta	
arterial pressure in foot	
venous pressure in foot	
capillary pressure in foot	
arterial pressure in cerebral blood vessels	

 STOP!   ANSWER ALL QUESTIONS ABOVE BEFORE PROCEEDING.

[1]The clinical problem presented here is based on J. E. Eisele, C. E. Cross, D. R. Rausch, C. J. Kurpershoek, and R. F. Zelis (1971), Abnormal respiratory control in acquired dysautonomia, *New England Journal of Medicine*, **285**, 366–368

A 47-year-old teacher goes to her physician complaining of dizziness, fainting, and shortness of breath. She has also experienced episodes of diarrhea.

    She appears pale and chronically ill. Her heart rate is 70/min supine (lying down) and 68/min sitting and standing. Her blood pressure is 120/74 mm Hg supine, 98/54 mm Hg sitting, and 84/48 mm Hg standing. Heart sounds and lung sounds are normal. Her abdominal exam and her general neurologic exam are both normal. Routine lab exams are normal.

2. Identify any abnormal data in the description of this patient.

 STOP!   ANSWER ALL QUESTIONS ABOVE BEFORE PROCEEDING.

Several studies were performed to evaluate the patient's circulatory control mechanisms.

   She was placed on a tilt table, which allows rapid changes in the patient's orientation. Her blood pressure while supine was 102/60 mm Hg. She was then tilted to a 45° heads-up position, and her blood pressure fell to 70/35 mm Hg. There was no change in her heart rate when tilted.

3. Evaluate the results of the tilt test. Is the patient's response normal? What possible mechanism(s) could account for this result?

STOP!   ANSWER ALL QUESTIONS ABOVE BEFORE PROCEEDING.

Failure to maintain a nearly constant blood pressure in changing posture from the supine to the erect is called orthostatic hypotension.

4. List all the functional locations within the cardiovascular system where pathology or dysfunction could lead to failure to activate the baroreceptor reflex upon standing.

STOP! ANSWER ALL QUESTIONS ABOVE BEFORE PROCEEDING.

Administration of 2 mg of atropine (an antagonist of muscarinic receptors) intravenously did not result in any change in heart rate.

5. Is this response normal? How can you account for this result?

Isoproterenol (an α-adrenergic agonist) administration resulted in her heart increasing from 60/min to 118/min.

Stimulated release of catecholamines from sympathetic nerve endings (via administration of tyramine and monoamine oxidase inhibitor) produced a rise in blood pressure.

6. Are these responses normal? What do they tell you?

7. Where in the system do you believe the patient's problem lies? Why?

You have just completed Problem 5 in Unit 5.
Please turn the page to begin work on the next problem.

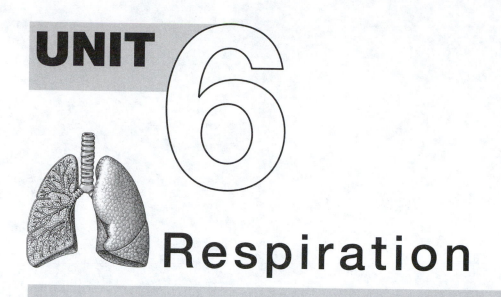

UNIT 6

Respiration

Problem 1

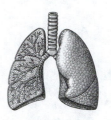

PROBLEM INTRODUCTION

The pattern of ventilation determines the alveolar ventilatory rate, an important determinant of alveolar and arterial gas composition. This problem provides data that may be used to determine the effects of various ventilatory patterns on alveolar gas partial pressures.

The following measurements were made on a normal subject.

PARAMETER	VALUE
P_aCO_2 (mm Hg)	36
P_aO_2 (mm Hg)	100
$\dot{V}_E$ (L/min)	9
$\dot{V}_A$ (L/min)	6
f (breaths/min)	15

DEFINITION OF SYMBOLS	
P_a	arterial partial pressure
$\dot{V}_E$	minute ventilation
$\dot{V}_A$	alveolar ventilation
V_T	tidal volume
V_D	dead space volume
f	breathing frequency

1. Which of the breathing pattern(s) below would not change the subject's P_aO_2? Explain your answer.

PATTERN	V_T (ml)	f (per min)
a	400	30
b	500	30
c	800	7.5
d	800	10
e	900	10

STOP!   ANSWER ALL QUESTIONS ABOVE BEFORE PROCEEDING.

2. Which breathing pattern(s) would reduce the subject's P_aCO_2 to 24 mm Hg? Explain your answer.

You have just completed Problem 1 in Unit 6.
Please turn the page to begin work on the next problem.

Problem 2

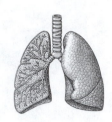

PROBLEM INTRODUCTION

Oxygen delivery to the tissues is essential for maintaining their viability and function. One component in this process is oxygen uptake into blood, a process that significantly depends on the oxygen transport protein, hemoglobin. This problem explores the consequences of a hemoglobin deficiency, anemia, on oxygen transport and delivery.

The questions below have to do with identical twins A and B. Twin A is normal, but twin B is anemic. Twin A has a hemoglobin concentration of 15 gm/dl, whereas twin B's is 7.5 gm/dl, one-half that of A. Assume that twin B's anemia has not caused any compensatory changes. An oxyhemoglobin dissociation curve is shown in Figure resp 2-1. Enter your results in Table 2-1.

Both twins have a P_AO_2 (alveolar O_2 partial pressure) of 100 mm Hg.

1. What is twin A's:

   a. P_aO_2 (arterial PO_2)?

   b. Arterial hemoglobin O_2 capacity ($capHbO_2$)?

   c. Arterial O_2 content (C_aO_2)?

TABLE 2-1

TABLE 1	P_aO_2 (mm Hg)	$capHbO_2$ (ml O_2/dl blood)	CaO_2 (ml O_2/dl blood)
Twin A			
Twin B			

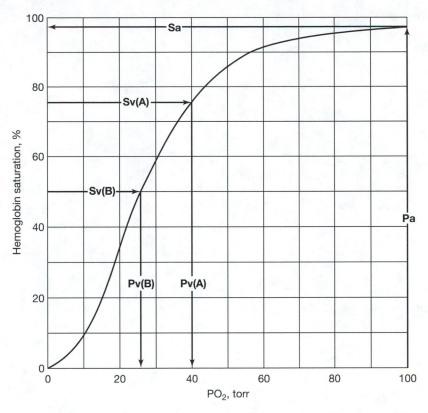

Figure resp 2-2. The oxyhemoglobin dissociation curve shows the relationship between oxygen partial pressure and hemoglobin saturation.

STOP! ANSWER ALL QUESTIONS ABOVE BEFORE PROCEEDING.

2. What is twin B's:

   a. PaO_2 (arterial PO_2)?

   b. Arterial hemoglobin O_2 capacity ($capHb_{O_2}$)?

   c. Arterial O_2 content (C_aO_2)?

   Enter your results in Table 2-1.

3. What value of arterial PO_2 would cause twin A's arterial O_2 content to be the same as twin B's?

4. Is it possible for hyperventilation to increase twin B's O_2 content to equal twin A's? Would breathing 100% O_2 do it?

 STOP!   ANSWER ALL QUESTIONS ABOVE BEFORE PROCEEDING.

At rest, both siblings have a cardiac output of 5 L/min and an O_2 consumption of 250 ml/min.

5. What is twin A's:
   a. C_vO_2?

   b. P_vO_2?

Enter your answers in Table 2-2.

TABLE 2-2

	C_vO_2 (ml O_2/dl blood)	P_vO_2 (mm Hg)
Twin A		
Twin B		

 STOP! ANSWER ALL QUESTIONS ABOVE BEFORE PROCEEDING.

6. What is twin B's:

   a. C_vO_2?

   b. P_vO_2?

   Enter your answers in Table 2-2.

7. What do these data tell you about the effects of anemia on oxygen diffusion from blood to the tissues?

You have just completed Problem 2 in Unit 6.
Please turn the page to begin work on the next problem.

Problem 3

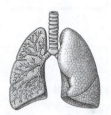

PROBLEM INTRODUCTION

Hypoxia is a condition in which the oxygen delivery to the tissues is reduced. There are a number of causes for hypoxia. Although each restricts oxygen delivery, each has a unique set of other characteristics.

1. A healthy college student went to Colorado at the end of the school year for a vacation. While there, he drove to the top of Pike's Peak, a tall mountain. During his first hour on the peak, he walked about but had to sit down to rest three times. The last time was in the weather station. He glanced at the barometer hanging over his head. It read 450 mm Hg.

   Predict the directional changes [increase (I), decrease (D), no change (0)] that would have taken place in each of the listed parameters by the time he entered the weather station compared with the values that he would have had at home (sea level).

	P_AO_2 (mm Hg)	P_aO_2 (mm Hg)	S_a (%)	S_v (%)	V_T (ml)	P_aCO_2 (mm Hg)	pH_a
Prediction							

DEFINITION OF SYMBOLS	
P	partial pressure
A	alveolar
a	arterial
S	Hb saturation
V_T	tidal volume
v	venous

 STOP! ANSWER ALL QUESTIONS ABOVE BEFORE PROCEEDING.

2. Predict the effects of the following in an otherwise normal individual at sea level.

	P_AO_2	P_aO_2	S_a	S_v	V_T	P_aCO_2	pH_a
Moderate exercise							
Pneumonia							
Partial tracheal obstruction							
CNS drug depression							
R to L shunt*							

*R to L shunt = blood flow from right side of heart to left side bypassing the lungs.

You have just completed Problem 3 in Unit 6.
Please turn the page to begin work on the next problem.

Problem 4

PROBLEM INTRODUCTION

The activity of the central and peripheral chemoreceptors determines the resting ventilation. This problem deals with a situation in which these chemoreceptor inputs have been modified by disease.

A 59-year-old carpenter came into the emergency room complaining of shortness of breath. His pulse was 112/min, and his blood pressure 138/88 mm Hg. His respiratory rate was 35/min, and his breathing was extremely labored. Loud breath sounds with some crackling rales were heard over all the lung fields. The patient had an ashen complexion, and his nail beds were cyanotic.

An arterial blood sample was drawn and sent to the lab for blood gas analysis. A chest X ray was ordered, and the patient was given oxygen to breathe.

1. Given the information you have about the patient, predict the deviations from normal (increase, decrease, no change) you would expect him to have in the following parameters *before he was given oxygen*. Explain your predictions.

PARAMETER	PREDICTION
P_aO_2 (mm Hg)	
P_aCO_2 (mm Hg)	
pH_a	
arterial O_2 content (ml/dl)	

STOP!   ANSWER ALL QUESTIONS ABOVE BEFORE PROCEEDING.

One-half hour later, the patient was found to be unresponsive. His complexion had changed to a flushed pink with no trace of cyanosis. His respiratory rate was 9/min, and his breathing was quiet. His heart rate was 140/min, and his blood pressure was 85/50 mm Hg. The patient was in a deep coma.

2. Compared with the time the patient came into the ER, predict the changes in the values of the parameters below (increase, decrease, no change) when he became comatose. Explain your predictions.

PARAMETER	PREDICTION
P_aO_2 (mm Hg)	
P_aCO_2 (mm Hg)	
pH_a	
arterial O_2 content (ml/dl)	

STOP!   ANSWER ALL QUESTIONS ABOVE BEFORE PROCEEDING.

The results of the blood-gas analysis from the blood sample that was drawn *at the time of his admission to the ER* were received and are given below.

3. Classify each of the laboratory measurements (high, low, normal).

MEASUREMENT	VALUE	HIGH, LOW, NORMAL
[Hb]	16 gm/dl	
hematocrit	52%	
P_aO_2	48 mm Hg	
P_aCO_2	90 mm Hg	
pH_a	7.35	
$[HCO_3^-]$	48 mM	

4. Has the patient had his "problem" for:
   a. one hour
   b. one day
   c. one week
   d. one year
   e. more than one year

5. What information did you use to arrive at your answer to the previous question?

STOP! ANSWER ALL QUESTIONS ABOVE BEFORE PROCEEDING.

6. When the patient was first admitted, his respiratory rate was 35/min. When he was comatose, however, his respiratory rate was 9/min. What caused the change?

7. Diagram the sequence of events to show why the patient became comatose.

You have just completed Problem 4 in Unit 6.
Please turn the page to begin work on the next problem.

Problem 5

PROBLEM INTRODUCTION

Physiological systems do not exist in isolation. A perturbation in one system is usually followed by a change in the function of other systems as well. Depending on the disturbance, the effects on other systems may be minimal or quite severe. This problem illustrates an interaction between the cardiovascular and respiratory systems.

Mrs. V.H. is a 45-year-old, overweight woman who was determined, using cardiac catheterization, to have a mitral valve problem (see Cardiovascular 4). She has complained of increasing shortness of breath over the past two to three weeks. She has "three pillow" orthopnea and swollen feet. She was admitted to the hospital having a cough productive of whitish sputum and a sharp retrosternal pain. Her respiration was labored, and she had diffuse wheezing.

Her chest X ray showed an enlarged left atrium, increased pulmonary vascular markings (increased filling of her pulmonary vessels), and an indication of interstitial pulmonary edema. Her electrocardiogram showed sinus tachycardia and a suggestion of right ventricular hypertrophy.

On examination her vital signs were:

VITAL SIGN	VALUE
heart rate	90/MIN
respiratory rate	26/MIN
blood pressure	115/70
temperature	99.2° F

1. List the abnormal findings in the patient's description that are indicative of a respiratory problem on the following page.

STOP!   ANSWER ALL QUESTIONS ABOVE BEFORE PROCEEDING.

ABNORMAL RESPIRATORY FINDINGS

---

2. The patient has a mitral valve stenosis (narrowing) and incompetence (does not close completely). Draw a causal diagram showing how this condition accounts for the signs and symptoms you listed above.

STOP!   ANSWER ALL QUESTIONS ABOVE BEFORE PROCEEDING.

As a consequence of her diseased heart valve, she has pulmonary hypertension.

3. Although it was felt that a major component of her shortness of breath was cardiac in origin, pulmonary function tests were performed to evaluate her for the presence of primary pulmonary disease. Assuming that she does *NOT* have primary lung disease, predict the changes (increased, decreased, unchanged from normal) you would expect to find in this patient in the parameters listed below.

MEASUREMENT		PREDICTION
Spirometry	vital capacity (L)	
	FEV_1	
	FEV_1/FVC (%)	
	MMEF (L/sec)	
Lung volumes	FRC (L)	
	RV (L)	
	TLC (L)	
Other	D_LCO (cc/min/mm Hg)	

STOP! ANSWER ALL QUESTIONS ABOVE BEFORE PROCEEDING.

Her measured pulmonary function data are shown below.

MEASUREMENT		OBSERVED	EXPECTED	OBSERVED/ EXPECTED (%)
Spirometry	Vital capacity (L)	3.11	5.09	61
	FEV$_1$ (L)	2.39	3.67	65
	FEV$_1$/FVC (%)	77	72	107
	MMEF (L/sec)	2.11	3.51	60
Lung volumes	FRC (L)	2.06	2.42	85
	RV (L)	0.99	1.24	80
	TLC (L)	4.10	5.94	69
other	D$_L$CO (cc/min/mmHg)	21.98	29.3	75

4. What change (increase, decrease, no change) would you expect to be present in each of the following?

PARAMETER	PREDICTION
arterial PO$_2$	
arterial Hb saturation	
arterial PCO$_2$	
arterial pH	
lung compliance	

You have just completed Problem 5 in Unit 7.
Please turn the page to begin work on the next problem.

UNIT 7

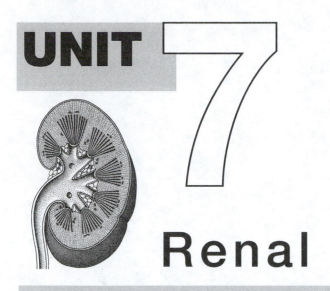

Renal

Problem 1

PROBLEM INTRODUCTION

Measurement of the clearance of substances from the blood by the kidneys provides one means of evaluating kidney function. Furthermore, the clearance of certain substances can be used to determine such important parameters as the glomerular filtration rate and the renal plasma flow. The renal clearances of other substances can provide insight into the role of the kidney in maintaining homeostasis.[1]

The data presented below were collected from a healthy student on three successive days. The data in the column "Ad Lib" were collected while the student was allowed to drink water as she chose. The data labelled "Thirsting" were collected after 12 hours of no fluid intake. The "+1 L Water" data were collected 90 minutes after she drank 1 liter of water. The numbers in the table below that are underlined have been calculated from the data given.

The table below contains the correct predictions.

PARAMETER	AD LIB	THIRSTING	+1 L WATER
V (ml/min)	1.2	0.75	15.0
U_{In} (mg/ml)	15.8	25.2	1.23
P_{In} (mg/ml)	0.151	0.155	0.154
GFR (ml/min)	<u>126</u>	?	?
% filtered H_2O reabsorbed	<u>99.0</u>	?	?
P_{Na} (mM)	136	144	134
L_{Na} (mmoles/min)	<u>17.1</u>	<u>17.6</u>	<u>16.1</u>
U_{Na} (mmoles/L)	128	192	10.2
% filtered Na reabsorbed	99.1	?	?

[1]This problem is based on Problem 8-1, page 141–143 of H. Valtin. (1967). *Renal Function: Mechanisims Preserving Fluid and Sollute Balance in Health.* Boston: Little, Brown, and Company with permission of the publishers.

PARAMETER	AD LIB	THIRSTING	+1 L WATER
U_{osm} (mOsm/L)	663	1000	100
P_{osm} (mOsm/L)	290	300	287
C_{H_2O} (ml/min)	−1.54	?	?
U_{urea} (mg/100 ml)	480	720	48
P_{urea} (mg/100 ml)	12	15	10
Curea (ml/min)	48	?	?
% filtered urea reabsorbed	61	?	?

1. Predict the changes relative to the ad lib state (increase,decrease,no change) that will be present in the data labeled with question marks (?) after Thirsting and after drinking 1 liter of water.

✋ **STOP! ANSWER ALL QUESTIONS ABOVE BEFORE PROCEEDING.**

2. Now calculate the unknown values (shown as question marks) using the data provided in the table.

3. Compare your predictions with the actual values you have calculated. Explain the changes that are present in the thirsting state and after drinking 1 liter of WATER.

☞ *You have just completed Problem 1 in Unit 7.*
Please turn the page to begin work on the next problem.

Problem 2

Consider an individual who develops a pathology that only results in a leaky glomerulus (permeable to plasma proteins).

1. Predict the changes (increase/decrease/no change) that will be present in the following parameters in the steady state as a consequence of this pathology.

PARAMETER	PREDICTED CHANGE
plasma protein concentration	
plasma colloid osmotic (oncotic) pressure	
total body sodium	
plasma volume	
interstitial fluid volume	

 STOP!   ANSWER ALL QUESTIONS ABOVE BEFORE PROCEEDING.

As a result of fluid shifts between the vascular and the interstitial compartments additional changes in function affect cardiovascular and renal function.

2. Predict the changes (increase, decrease, no change) that will occur in the following parameters.

PARAMETER	PREDICTED CHANGE
mean arterial pressure	
glomerular filtration rate	
renal blood flow	
plasma renin concentration	
plasma aldosterone concentration	
Na excretion	
ADH secretion	

STOP! ANSWER ALL QUESTIONS ABOVE BEFORE PROCEEDING.

An infection with group A streptococcus ("strep throat") can be followed by kidney damage characterized by the development of a leaky glomerulus (increased permeability to plasma proteins) and damage to some renal tubules *and* destruction of others.

3. Predict the <u>initial</u> qualitative changes (increase, decrease, no change) that will occur in the following parameters as a result of this pathology.

PARAMETER	PREDICTED CHANGE
glomerular filtration rate	
plasma protein concentration	
plasma colloid osmotic (oncotic) pressure	
total body sodium	
plasma volume	
interstitial fluid volume	

STOP!   ANSWER ALL QUESTIONS ABOVE BEFORE PROCEEDING.

As a result of the pathology present and the compensatory mechanisms that are activated, there are additional changes to body function.

4. Predict the changes (increase, decrease, no change) that will occur in the following parameters.

PARAMETER	PREDICTED CHANGE
mean arterial pressure	
plasma renin concentration	
plasma aldosterone concentration	
Na excretion	
ADH secretion	

STOP! ANSWER ALL QUESTIONS ABOVE BEFORE PROCEEDING.

5. In the first patient the consequences of a leaky glomerulus leads to a sustained *decrease* in blood pressure. In the second case, in which leaky glomerulus and tubular damage are present, the result is a sustained *increase* in blood pressure. Explain these two quite different outcomes of similar (but, of course, not identical) pathology.

You have just completed Problem 2 in Unit 7.
Please turn the page to begin work on the next problem.

Problem 3

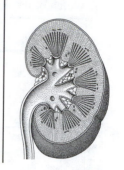

PROBLEM INTRODUCTION

Abnormal function in one organ or organ system rarely remains isolated to that system. This statement is particularly true when the abnormality lies in some component of the nervous or endocrine systems; these systems control the function of all the other organ systems. Disturbances to water and electrolyte balance, initiated by abnormal endocrine function and acting via altered function in the kidney, ultimately can cause cardiovascular problems as well.

Ms. R.C.V. is a 45-year-old Caucasian housewife. She has come to her physician complaining of extreme weakness. Her last visit was 5 years ago; at that time her physician found that she had a mild hypertension and recommended that she restrict her salt intake.

Her physician took a history and did a physical examination, the results of which are provided later. Blood and urine samples were also taken for analysis. The results of these are given below.

MEASUREMENT	PLASMA VALUE (normal range)	URINE VALUE
Na^+	145 mEq/L	
K^+	2.5 mEq/L	
HCO_3^-	37 mEq/L	0
Cl^-	98 mEq/L	
pH	7.53	6.9
BUN	22 mg/dl (15-39)	
creatinine	0.71 mg/dl (0.53-1.19)	
protein	7.5 gm/dl (6.5-8.0)	negative

BUN = blood urea nitrogen

1. Which of the plasma and urine values are abnormal?

	Plasma	Urine
Na+		
K+		
HCO$_3^-$		
Cl$^-$		
pH		
BUN		
creatinine		
protein		

2. What single pathology or pathophysiological condition could account for her abnormal plasma and urine values?

STOP!   ANSWER ALL QUESTIONS ABOVE BEFORE PROCEEDING.

On the basis of the data available, the attending physician requested some special blood tests. They revealed that the patient's blood aldosterone level was *elevated* and that her plasma renin was *lower* than normal. The physician concluded that the patient most likely had an aldosterone hypersecreting adrenal tumor.

3. Predict the consequence I, D, O of hyperaldosteronism on the following:

	RENAL EXCRETION	PLASMA CONCENTRATION
Na+		
K+		
H+		
HCO$_3$$^-$		

4. Draw a concept map that explains your predictions.

STOP! ANSWER ALL QUESTIONS ABOVE BEFORE PROCEEDING.

The patient reports that she has gained some weight, which she attributes to getting older. She says that she has a tendency toward puffiness that she thinks is related to her menstrual cycle.

Physical examination revealed a woman with mild edema in her lower extremities. Her blood pressure was 150/95 and did not change after a period of rest. Ms. R.C.V. had complained about muscular weakness; indeed, her physician found her to be moving sluggishly and her reflexes were distinctly slowed.

Her ECG was taken and revealed ST depression, a flattened T wave, and distinct U waves. A rhythm strip (a long duration recording) showed occasional ventricular arrhythmias. All these electrocardio-graphic abnormalities are commony seen with hypokelemia.

5. Identify the abnormalities in the patients history and physical.

6. Expand your concept map to explain these abnormalities, assuming that they all result from her endocrine condition.

Problem 4

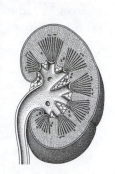

PROBLEM INTRODUCTION

The kidneys play a major role in regulating the composition of body fluids. They carry out this function by transporting substances between the vascular compartment and the lumen of the nephron. An understanding of these transport processes and their controls is essential to understand the function of the kidneys as a homeostatic regulator.

Some of the substances present in the glomerular filtrate are reabsorbed and are returned to the body, whereas other substances present in the vascular compartment are secreted into the tubular lumen.

1. What routes are available for the movement of substances between the tubular lumen and the peritubular extracellular fluid?

 STOP! ANSWER ALL QUESTIONS ABOVE BEFORE PROCEEDING.

Substances can move between the tubular lumen and the extracellular compartment by two pathways: (1) a *paracellular* pathway between two adjacent cells making up the tubule and (2) a *transcellular* pathway across the walls of and through the tubular cells.

2. What transport processes move a substance through the paracellular pathway? What are some substances that are transported through the paracellular pathway?

3. What transport processes are involved in moving substances across the wall of the nephron by the intracellular pathway? What are some substances that are transported via the transcellular pathway?

STOP!   ANSWER ALL QUESTIONS ABOVE BEFORE PROCEEDING.

4. Give one example of a physiologically important substance that is reabsorbed or secreted by *primary active transport*. Describe the mechanism by which this substance is transported between the tubular fluid and the extracellular fluid.

5. Give one example of a physiologically important substance that is reabsorbed or secreted by *secondary active transport*. Describe the mechanism by which this substance is transported. What determines the maximum rate at which this solute can be reabsorbed or secreted by this transport process? Of what significance is this transport rate limitation to the function of the kidney?

6. Give one example of a substance that is reabsorbed or secreted *passively*. Describe the mechanism of this transport. What forces are present to drive this transport?

STOP! ANSWER ALL QUESTIONS ABOVE BEFORE PROCEEDING.

7. Describe the handling of the following substances at each segment of the nephron by indicating for each segment whether net reabsorption (R) or net secretion (S) occurs there.

	NA+	K+	GLUCOSE	UREA	WATER
proximal tubule					
descending limb of loop of Henle					
ascending limb of loop of Henle					
distal tubule					
collecting duct					

8. Transport of some of these substances is subject to physiological controls via hormones. In the table below, indicate which substances have transport processes that are controlled by hormones, where in the kidney this control is exerted, and what hormones are involved.

	NA+	K+	GLUCOSE	UREA	WATER
proximal tubule					
descending limb of loop of Henle					
ascending limb of loop of Henle					
distal tubule					
collecting duct					

9. List some other substances whose reabsorption in the kidney is under endocrine control.

You have just completed Problem 4 in Unit 7.
Please turn the page to begin work on the next problem.

Problem 5

1. Acetazolamide is a drug that inhibits the enzyme carbonic anhydrase in the cells of the proximal tubule. Predict the effects (increase/decrease/no change) of acetazolamide on the following parameters.

	URINE CONCENTRATION OF				URINE FLOW RATE
	Na^+	K^+	Cl^-	HCO_3^-	
acetazolamide					

2. Explain the mechanism by which acetazolamide produces these effects.

 STOP! ANSWER ALL QUESTIONS ABOVE BEFORE PROCEEDING.

3. <u>Furosemide</u> is a drug that inhibits the active reabsorption of solute in the ascending limb of the loop of Henle by blocking the Na, K, 2Cl co-transporter in the luminal membrane. Predict (increase/decrease/no change) the effects of furosemide on the following parameters.

	URINE CONCENTRATION OF				URINE FLOW RATE
	Na^+	K^+	Cl^-	HCO_3^-	
acetazolamide					
furosemide					

4. Explain the mechanism by which furosemide produces these effects.

STOP! ANSWER ALL QUESTIONS ABOVE BEFORE PROCEEDING.

5. Thiazide is a drug that decreases the reabsorption of Na in the early portion of the distal tubule. Predict the effects of thiazide (increase/decrease/no change) on each of the following parameters.

	URINE CONCENTRATION OF				URINE FLOW RATE
	Na+	K+	Cl–	HCO₃–	
acetazolamide					
furosemide					
thiazide					

6. Explain the mechanism by which thiazide produces these effects.

STOP!   ANSWER ALL QUESTIONS ABOVE BEFORE PROCEEDING.

7. Spironolone is a drug that blocks the action of aldosterone on the collecting duct. Predict the effects (increase/decrease/no change) of spironolone on the following parameters.

	URINE CONCENTRATION OF				URINE FLOW RATE
	Na$^+$	K$^+$	Cl$^-$	HCO$_3^-$	
acetazolamide					
furosemide					
thiazide					
spironolone					

8. Explain the mechanism by which spironolone produces these effects.

STOP!   ANSWER ALL QUESTIONS ABOVE BEFORE PROCEEDING.

9. Triamterene is a drug that blocks Na reabsorption in the collecting duct by blocking the pumping of Na into the cell. Predict the effects of triamterene (increase/decrease/no change) on the following parameters.

	URINE CONCENTRATION OF				URINE FLOW RATE
	Na^+	K^+	Cl^-	HCO_3^-	
acetazolamide					
furosemide					
thiazide					
spironolone					
triamterene					

10. Explain the mechanism by which triamterene produces these effects.

11. Diuretics are not uncommonly self-prescribed to achieve weight loss for cosmetic or sports-related reasons. What homeostatic imbalances might result from such medically uncontrolled use of these drugs?

12. What is the role of active transport in enabling the kidneys to produce a urine that is more concentrated, or less concentrated, than body fluids?

You have just completed Problem 5 in Unit 7.
Please turn the page to begin work on the next problem.

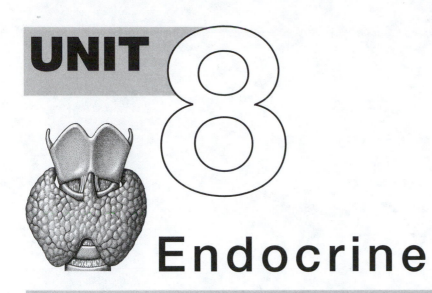

8

Endocrine

Problem 1

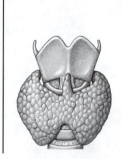

PROBLEM INTRODUCTION

All the endocrine systems are organized in a manner that permits them to carry out their function: the control of bodily activity in the service of homeostasis. Such systems respond to disturbances so as to minimize the effect of these disturbances on the system's regulated variables.

Figure endo 1-1, p. 138 is a diagram of a hypothetical endocrine control system. The following abbreviations are used here and in the figure:

H = the hormone product of the endocrine cell

C = a specific plasma carrier of hormone H

X = a normal stimulus to the endocrine cell

The lowercase letters in the figure refer to situations defined below. Determine the consequences of each situation on the hormone concentration in the blood ([H]) and the hormone EFFECT. As used here, EFFECT is a gross action of a hormone. For example, an EFFECT of insulin is to reduce the plasma glucose concentration. Assume that the condition described in each situation is maintained for a long time. Predict the changes you expect to occur in each situation in two time periods: (1) the immediate consequences and (2) the steady-state consequences. Use $++$ for a large increase, $+$ for a smaller one; $--$ for a large decrease, $-$ for a smaller one; and 0 for no change or a return to the initial condition.

 STOP! ANSWER ALL QUESTIONS ABOVE BEFORE PROCEEDING.

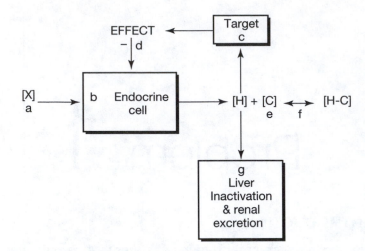

Figure endo1-1. A hypothetical endocrine control system that regulates the plasma concentration of the hormone H ([H]). The lowercase letters refer to various situations that act as disturbances to the system. See the text for a definition of situations.

SITUATION	IMMEDIATE CONSEQUENCES		STEADY-STATE	
	[H]	EFFECT	[H]	EFFECT
a				
b				
c				
d				
e				
f				
g				

Situations:

a. [X] is increased. Draw a graph of [H] versus time starting before the increase in [X] and continuing into the new steady state.

b. The biochemical pathways that enable X to stimulate H secretion are partially inhibited.

c. The number of H receptors is decreased.

d. The endocrine cell response to the EFFECT is facilitated

e. The plasma carrier concentration, [C], is suddenly increased. Draw a graph of [H] and [HC] on the same time axis starting before the increase in [C] and continuing to the new steady state.

f. The hormone binding constant of C is decreased.

g. Liver or renal failure leads to a decrease in H inactivation or excretion.

You have just completed Problem 1 in Unit 8.
Please turn the page to begin work on the next problem.

Problem 2

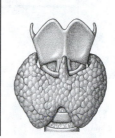

PROBLEM INTRODUCTION

The anterior pituitary gland secretes many hormones, some of which are trophic to peripheral endocrine glands. Changes in the secretion rate of any of these hormones produce characteristic effects. This problem deals with a situation in which a patient experiences changes in hormone secretion.

Medical history

A 51-year-old man came to the emergency room. He complained of 4 days of a retrobulbar headache of increasing severity. He also reported 8 years of decreased libido, 3 years of chronic fatigue, cold intolerance, decreased beard growth, and poor tanning of the skin. He has had a progressive increase in shoe and ring size over the past 20 years, along with increasing enlargement of his nose, lips, tongue, and jaw.

Physical examination

The patient's blood pressure (BP) was 100/60 mm Hg lying and 80/50 mm Hg standing. His skin was dry and pale, and it appeared thickened. Pubic and axillary hair were scanty. His testes were smaller and softer than normal. Deep tendon reflexes showed a prolonged relaxation phase. His nose, jaw, zygomata, and lips were prominent; his tongue was large, and the soft tissue of his hands and feet were enlarged. He exhibited bitemporal hemianopsia (blindness in his lateral visual fields); his left pupil was dilated and his left eye deviated laterally.

1. In the table below, list the findings in the patient history and physical that seem to be abnormal. Collect these into groups, each of which seems to you to be associated with a single hormone system. Identify the hormone associated with each group and indicate whether the data suggest an increase or a decrease (I or D) in hormone secretion rate.

STOP! ANSWER ALL QUESTIONS ABOVE BEFORE PROCEEDING.

HORMONE	I/D	ABNORMALITIES

 STOP!   ANSWER ALL QUESTIONS ABOVE BEFORE PROCEEDING.

Initial Lab Workup

Laboratory studies were ordered with the following results.

TEST	RESULTS
urinalysis	unremarkable
white blood count	6000 with 6% eosinophils (normal <3%)
serum sodium	128 mEq/L
serum potassium	4.7 mEq/L
plasma cortisol	3μg/dl (normal 7-25 at 8:00 a.m.)

2. Do these results confirm or deny any of your hypothesized hormone imbalances? Explain how you used the data to confirm or deny your hypotheses.

STOP!   ANSWER ALL QUESTIONS ABOVE BEFORE PROCEEDING.

Over the next 8 hours, the patient was given fluids and hydrocortisone intravenously. His supine BP increased to 120/80 mm Hg, his headache worsened, and his condition obviously deteriorated.

3. Explain what happened to the patient during this period.

STOP!   ANSWER ALL QUESTIONS ABOVE BEFORE PROCEEDING.

Endocrine Lab Workup

The results of other laboratory studies have just been received. (The specimens were drawn on admission, but results have just been returned.)

TEST	RESULTS
serum thyroxine	3.1 µg/dl (normal 4–11)
serum TSH	<1µU/ml (normal 0–10)
serum testosterone	189 ng/100 ml (normal male 300–1100)
serum LH	1.7 mIU/ml (normal 1–15)
serum PRL	4 ng/ml (normal male <15)
serum GH	47 ng/ml (normal <5)

4. What is your final diagnosis? Explain your reasoning.

STOP!   ANSWER ALL QUESTIONS ABOVE BEFORE PROCEEDING.

The patient is suffering from a growth hormone–secreting tumor of his anterior pituitary, acromegaly, of 20 years duration. In addtion, several features of hypopituitarism have been present for at least 8 years.

5. Draw a concept map relating the patient's pathology to his signs and symptoms.

6. How would you treat this patient? What long-term consequences would you have to deal with, and how would you handle them?

You have just completed Problem 2 in Unit 8.
Please turn the page to begin work on the next problem.

Problem 3

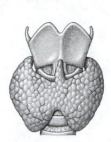

A mother brought her 13-year-old daughter, Jenny, to her family physician. The mother was concerned about what she described as "emotional disturbances" in the child (bed-wetting and generalized "laziness"). Upon questioning, the daughter revealed that she had recently developed a habit of eating between meals and drinking a great deal of water to satisfy her thirst. She had started having menstrual periods 6 months earlier.

1. Define two possible pathophysiological mechanisms that could give rise to the symptoms described in this history.

 STOP! ANSWER ALL QUESTIONS ABOVE BEFORE PROCEEDING.

The patient appeared to be a rather thin but otherwise normal adolescent. She did, however, have glucosuria (sugar in the urine) and a blood sugar of 350 mg/dl.

2. What disease does this patient have?

STOP!   ANSWER ALL QUESTIONS ABOVE BEFORE PROCEEDING.

The patient is diabetic. Her blood sugar was brought under control with daily insulin injections and has been well controlled since then.

At age 16 (3 years after her first hospitalization), the patient was admitted to the emergency room in a coma. The mother reported that her daughter had attended a school dance the previous evening and had arrived home very late. She was allowed to sleep the next morning and no attempt had been made to wake her until well after noon. At that time she could not be adequately aroused.

Examination revealed a comatose patient with slow but very deep respirations. Her urine was positive for glucose and ketone bodies.

3. Why has the patient become ill?

 STOP!   ANSWER ALL QUESTIONS ABOVE BEFORE PROCEEDING.

Jenny's blood chemistries from a sample drawn on admission to the ER were the following:

BLOOD CHEMISTRY	VALUE
Na^+	152 mM
K^+	7.1 mM
Cl^-	88 mM
HCO_3^-	10.8 mM
pH	7.27
BUN	35 mg/dl

4. Explain each of the patient's signs, symptoms, and lab values while she was comatose.

STOP!   ANSWER ALL QUESTIONS ABOVE BEFORE PROCEEDING.

The patient was treated with insulin and intravenous saline. Initially, she showed some improvement, but 2 hours later she again became stuporous. At that time, she had a blood pressure of 80/50 mm Hg, and her blood chemistries were the following.

BLOOD CHEMISTRY	VALUE
glucose	100 mg/dl
ketone bodies	absent
pH	7.38
K+	2.2 mM

5. Explain why she again became comatose.

You have just completed Problem 3 in Unit 8.
Please turn the page to begin work on the next problem.

Problem 4

PROBLEM INTRODUCTION

Endocrine glands are components of control systems which function by negative feedback. These systems oppose the actions of disturbances that would tend to disrupt the normal functions that the system regulates. This problem provides an example of such interactions.

Five years ago, Ms. T.H., a 32-year-old secretary, was referred to you by her gynecologist. The referring physician had found an enlarged thyroid during a routine examination. Upon questioning, the patient reported none of the symptoms of thyroid disease. She did say, however, that both her sister and her mother took thyroid medication, but for unknown reasons. Physical examination revealed that her thyroid was indeed about twice normal size.

Blood samples were drawn for the measurement of TSH and free T_4.

1. Predict the outcome of these measurements relative to normal (I = greater, D = lower, 0 = normal).

HORMONE	VALUE
TSH	
free T_4	

STOP! ANSWER ALL QUESTIONS ABOVE BEFORE PROCEEDING.

The results of the plasma hormone study are shown below.

HORMONE	VALUE
TSH	4.1 mU/L (normal 0.4–4.2)
free T_4	0.9 ng/dl (normal 0.8–2.7)

2. What is the cause of her enlarged thyroid gland?

STOP!   ANSWER ALL QUESTIONS ABOVE BEFORE PROCEEDING.

It is now 5 years later, and Ms. T.H. has come to see you again. You have not seen her in the intervening period. She now complains of a 10-pound weight gain. She also complains of tiredness, sensitivity to cold, and heavier menses than usual. On examination, you find that her thyroid gland is about three times normal size. Her skin is somewhat dry. Her pulse is 56, lower than at her last examination.

A blood sample was drawn for hormone measurements.

3. Predict the outcome, relative to the measurements made five years ago (I = higher than before, D = lower than before, 0 = the same as before) of her plasma hormone concentrations. Explain the basis for your predictions.

HORMONE	VALUE
TSH	
free T_4	

STOP! *ANSWER ALL QUESTIONS ABOVE BEFORE PROCEEDING.*

The results of her hormone study are given below.

HORMONE	VALUE
TSH	18.6 mU/L (normal 0.4–4.2)
free T$_4$	0.6 mg/dl (normal 0.8–2.7)

In addition, her antithyroid peroxidase antibodies were determined to be 142.6 IU/ml (normal <2.0).

4. What is the cause of Ms. T.H.'s thyroid problem?

5. Why is she now exhibiting the symptoms of hypothyroidism?

6. Why has her thyroid size increased still further?

7. Draw a concept map showing the causal sequence of Ms. T.H.'s medical history.

☞ ***You have just completed Problem 4 in Unit 8.***
Please turn the page to begin work on the next problem.

Problem 5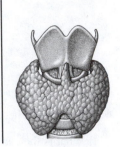

PROBLEM INTRODUCTION

A patient often exhibits signs and symptoms that could have resulted from any of a number of causes. To determine the specific cause that is operating in a particular case requires that the student know the possible causes and know what to measure that will allow these causes to be differentiated. The subject in this problem exhibits an assemblage of symptoms that could arise from a variety of causes.

Ms. A.C. is a 44-year old lawyer who was referred to a physician by her dentist, who noted hyperpigmentation of her gums. In any case, she had been meaning to see a doctor because of weakness, fatigue, and lightheadedness. She also stated that she had read an article in a magazine that convinced her that she was suffering from hypoglycemia: she develops a rapid heartbeat, shakiness, sweating, and faintness if she does not eat promptly on arising in the morning or if she misses a meal. These symptoms are relieved 15 to 20 minutes after eating. It is January in Chicago, yet the patient, who has not traveled, looks as though she just returned from a vacation in the Caribbean.

1. What are the possible causes of "lightheadedness" or "faintness"?

2. What are the possible hormonal and metabolic causes of hypoglycemia?

 STOP!   ANSWER ALL QUESTIONS ABOVE BEFORE PROCEEDING.

3. What could be the cause of Ms. A.C.'s symptoms upon missing a meal?

4. What pathophysiological mechanism(s) might be responsible for the signs and symptoms exhibited by Ms. A.C.? What data in the patient's description lead you to this hypothesis (these hypotheses)?

5. What would you do to test your hypothesis (hypotheses)?

STOP!   ANSWER ALL QUESTIONS ABOVE BEFORE PROCEEDING.

A standard blood workup was done. The table below shows the significant data.

TEST	RESULT
Na+	132 mM (normal 135–145)
K+	5.2 mM (normal 3.5–5.0)
fasting blood glucose	60 mg/dl (normal 70–120)

6. Do these data support or refute your hypothesis (hypotheses) about what is wrong with Ms. A.C.? What leads you to that conclusion?

7. What would you do to test your hypothesis (hypotheses)?

STOP!   ANSWER ALL QUESTIONS ABOVE BEFORE PROCEEDING.

The following additional laboratory tests were performed.

TEST	RESULT
morning ACTH	148 pg/ml (normal 8–52)
morning cortisol	6μg/dl (normal 10–20)

Cortisol was unchanged 1 hour after injecting 0.25 mg of synthetic 1–24 ACTH (normal 2 × increase).

8. Why were the hormone measurements made in the morning?

9. What's wrong with Ms. A.C.?

10. Draw a concept map that shows the causal relationships between Ms. A.C.'s disorder and her signs, symptoms, and laboratory values.

You have just completed Problem 5 in Unit 8.
Please turn the page to begin work on the next problem.

Problem 6

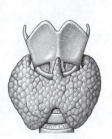

PROBLEM INTRODUCTION

When the heart and blood vessels are stimulated, the blood pressure increases. There are many possible ways in which this can be brought about. This problem deals with a situation in which a subject's condition causes her blood pressure to be labile.

A 30-year-old respiratory therapist went to her doctor asking for information about high blood pressure. She had been told 3 years ago, at the time of her pre-employment examination, that she had mild hypertension. She returned several times for follow-up checks. Her blood pressure was normal on three occasions and moderately elevated on a fourth visit. She was informed that she had "labile" hypertension (i.e., her hypertension was only intermittently increased) and was encouraged to have her blood pressure checked every 6 months or so.

1. What pathophysiological—abnormal physiological—mechanisms could produce hypertension on some occasions and not at others just a few weeks later?

 STOP! ANSWER ALL QUESTIONS ABOVE BEFORE PROCEEDING.

She had no further problem until 6 months ago. Then, she began to experience "spells" during which she became nauseous (sometimes to the point of vomiting), had a headache, and exhibited nervousness, sweating, and a rapid pulse. A fellow therapist took her blood pressure during several of these "spells." It was usually elevated. On one occasion her blood pressure was 200/140 mm Hg.

2. Can any of the pathophysiological mechanisms you previously identified give rise to the additional signs and symptoms that appear during her "spells"? Which ones? Explain how each could lead to these signs and symptoms.

 STOP!   ANSWER ALL QUESTIONS ABOVE BEFORE PROCEEDING.

The physician decided that the patient needed further evaluation. Her previous medical history was unremarkable, as were the histories of her immediate family members.

On examination she was found to be a normally developed white female in no obvious distress. Her BP was 158/95 mm Hg standing and 160/98 mm Hg sitting. Her pulse was 90/min, and her respiratory rate 16/min. Her temperature was 98.6°F.

The rest of her physical examination was unremarkable. SMA6 and SMA12 (screening blood chemistry tests), complete blood count (CBC, of red and white cells) and urine exams were normal, as were her ECG and chest X ray.

3. Does this additional information enable you to eliminate any of the possible pathophysiological mechanisms you had previously identified? What do you think is wrong with this patient?

STOP!   ANSWER ALL QUESTIONS ABOVE BEFORE PROCEEDING.

Because of the patient's history of "spells," it was decided to screen her for a pheochromocytoma (a hypersecreting tumor of the adrenal medulla), although only 1% of patients with hypertension have the disease for this reason.

4. What could you do to test for the presence of pheochromocytoma? What would you expect to see?

STOP!   ANSWER ALL QUESTIONS ABOVE BEFORE PROCEEDING.

The patient was instructed to collect her urine for 24 hours. The test results are shown below.

AGENT	TEST RESULT	NORMAL
vanillylmandelic acid (VMA)	11.3 mg	<7.0
metanephrine	5.1 mg	<1.5
catecholamines	632.0 mcg	<135
homovanillic acid	4.3 mg	<10

Computerized tomography of her abdomen revealed bilateral masses in the adrenals, and tumors were removed from both adrenals. Histological examination of the specimens were consistent with pheochromocytoma.

5. Draw a concept map explaining the source of her signs and symptoms.

You have just completed Problem 6 in Unit 8.
Please turn the page to begin work on the next problem.

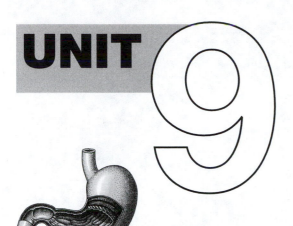

Gastrointestinal

Problem 1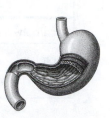

PROBLEM INTRODUCTION

The pancreas plays a key role in digestion and in controlling the body's fuel metabolism. Therefore, changes in either the exocrine or the endocrine functions of the pancreas seriously impact homeostasis.

Mr. G.I. is an alcoholic with a long history of chronic pancreatitis. During his most recent hospitalization, X rays taken showed extensive calcified fibrous tissue throughout the pancreas.

1. The *exocrine* pancreas is made up of two structures, each of which contributes to the production of pancreatic juice. List the two structural components and their secretory products in the table below.

STRUCTURES	SECRETORY PRODUCT(S)

STOP! ANSWER ALL QUESTIONS ABOVE BEFORE PROCEEDING.

2. Predict the consequences (increase, decrease, no change) of the loss of pancreatic acinar cells on the digestion and absorption of the following.

	DIGESTION	ABSORPTION
carbohydrates		
proteins		
fats		

3. Explain each prediction.

4. Predict the consequences (increase, decrease, no change) of the loss of pancreatic ductal tissue on the digestion and absorption of the following.

	DIGESTION	ABSORPTION
carbohydrates		
proteins		
fats		

5. Explain each prediction.

STOP!   ANSWER ALL QUESTIONS ABOVE BEFORE PROCEEDING.

The diagnosis of chronic pancreatitis is often made by delivering physiological stimuli to the pancreas by intravenous administration of some agent and then measuring the response using a catheter placed in the duodenum to collect samples of pancreatic juice.

6. What agent would you use to stimulate enzyme secretion? Explain.

7. What agent would you use to stimulate fluid secretion? Explain.

STOP! ANSWER ALL QUESTIONS ABOVE BEFORE PROCEEDING.

The pancreas also exhibits important *endocrine* functions. Pancreatitis can also affect these functions.

8. How might Mr. G.I.'s general metabolism be altered by a loss of pancreatic endocrine function?

You have just completed Problem 1 in Unit 9.
Please turn the page to begin work on the next problem.

Problem 2

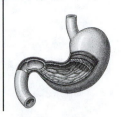

PROBLEM INTRODUCTION

The sequence of events that make up normal digestion requires the controlled movement of the food being processed through the gastrointestinal tract. If the timing of these events is disrupted, the consequences can be grave.

Mrs. Bilious is a 50-year-old woman with a history of juvenile onset arthritis. Management of her condition has included the use of large doses of aspirin and nonsteroidal anti-inflammatory drugs (NSAIDs). Recently she developed a gastric perforation that required removal of most of the body and all the antrum of the stomach (subtotal gastrectomy) and surgical intervention to facilitate gastric emptying (pyloroplasty, or surgical widening of the pylorus, and attachment of the remaining stomach directly to the jejunum).

1. Why did Mrs. Bilious develop a gastric ulcer that eventually perforated?

2. What effect (increase, decrease, no change) will the surgical procedures she has undergone have on the rate of gastric emptying? Explain.

 STOP! ANSWER ALL QUESTIONS ABOVE BEFORE PROCEEDING.

Gastric emptying will occur much more rapidly than normal, producing what is known as a "dumping" syndrome. Early symptoms of this condition include GI distress and vasomotor disturbances.

3. What will be the consequences of "dumping" on the functions of the duodenum? Explain.

4. What possible cardiovascular symptoms might be a part of the "dumping" syndrome? Explain.

5. How might the patient's problem be managed most simply?

You have just completed Problem 2 in Unit 9.
Please turn the page to begin work on the next problem.

Problem 3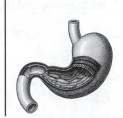

PROBLEM INTRODUCTION

The structure of the intestinal mucosa, both at the histological level and at the level of gross structure, is important for intestinal function. Abnormalities of structure will therefore result in altered intestinal function and disturbances to whole body homeostasis.

The mucosa of the small intestine is a heavily folded structure. It consists of villi projecting into the intestinal lumen, lined with the epithelial cells. The epithelia have microvilli on the luminal surface. Figure gi 3-1 illustrates the normal structure of the intestinal mucosa.

1. What two functions does intestinal mucosa serve in the overall process by which ingested food is made available to the cells of the body?

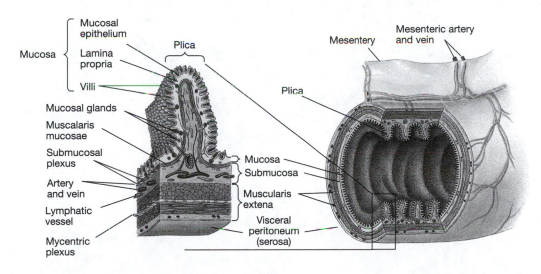

Figure gi 3-1. The normal structure of the intestinal mucosa (from Martini, *Fundamentals of Anatomy and Physiology*, 4th ed. Prentice Hall, 1998 p.864.).

 STOP! ANSWER ALL QUESTIONS ABOVE BEFORE PROCEEDING.

Pathology of the intestinal mucosa alters intestinal function; both digestion (the breaking down of complex molecules into smaller molecules) and absorption into the blood will be abnormal. Consider an individual whose intestinal villi have been altered by a tissue reaction to something ingested so that villi are flattened and the microvilli are shortened and blunted.

2. How will this pathology affect intestinal function? Explain.

3. Will all nutrients be equally affected? If not, what will be affected most? Why? Explain.

STOP! ANSWER ALL QUESTIONS ABOVE BEFORE PROCEEDING.

Gluten-sensitive enteropathy (sprue) is a condition in which the intestinal mucosa is damaged by a response to one of the molecular components of gluten. Sprue is one example of a malabsorption disease.

4. Predict the signs and symptoms of a sprue patient.

STOP! ANSWER ALL QUESTIONS ABOVE BEFORE PROCEEDING.

5. Explain the cause of each of the following signs and symptoms of a sprue patient.

SIGNS/SYMPTOMS	CAUSE
high fat content in stool	
diarrhea	
weight loss	
weakness and fatiguability	
anemia	
tetany and parasthesias	

You have just completed Problem 3 in Unit 9.
Please turn the page to begin work on the next problem.

Problem 4

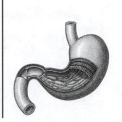

PROBLEM INTRODUCTION

The enzymes that work in the stomach require a very acidic pH for their optimum functioning. Such a low pH, however, is potentially very damaging to tissue, so the stomach (and the small intestine) must be protected from the dangerously low pH that is normally present in the stomach following the ingestion of a meal. The nature of the protective mechanisms involved and the consequences of disrupting this protection are considered here.

While digesting a meal, the pH of the stomach can get as low as 2.0.

1. Describe the mechanism by which the stomach secretes acid and achieves a pH of 2.0.

2. What stimuli control the secretion of acid in the stomach?

3. How is the stomach itself protected from the very acidic conditions that are present?

4. How is the duodenum protected from the acidity of the chyme that enters it as the stomach empties?

 STOP!   ANSWER ALL QUESTIONS ABOVE BEFORE PROCEEDING.

A gastric ulcer is a pathology in which conditions in the stomach result in cell damage or death and hence the erosion of the wall. The blood vessels of the stomach wall will also be damaged, and bleeding can thus result.

5. By what mechanism can an ulcer result from:
   a. prolonged and/or high doses of aspirin?

   b. hypersecretion of gastrin?

   c. stress?

6. Why is it fair to say that an active ulcer is an example of a positive feedback situation?

STOP!   ANSWER ALL QUESTIONS ABOVE BEFORE PROCEEDING.

Patients with Zollinger-Ellison syndrome (which is the result of a tumor of gastrin-secreting cells) exhibit (1) elevated serum gastrin concentrations, (2) increased basal secretion of stomach acid, and (3) peptic ulcer disease and/or diarrhea. Some patients also exhibit steatorrhea (fat in their stool) and evidence of malabsorption.

7. Explain the occurrence of steatorrhea and malabsorption in a Zollinger-Ellison patient.

You have just completed Problem 4 in Unit 9.
Please turn the page to begin work on the next problem.

Problem 5

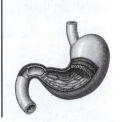

PROBLEM INTRODUCTION

Processing a mixed meal involves a complex sequence of events occurring over a relatively long time at a variety of sites along the gastrointestinal tract. The nature of each step and the processes that control and coordinate each one must be understood.

You go into your favorite Italian restaurant and order a meal of (A) spaghetti and (B) meatballs in (C) marinara sauce with a (D) nondiet soft drink.

1. Describe the predominant biochemical constituents of each of the components of this meal.

MEAL	CHEMICAL CONSTITUENTS
spaghetti	
meatballs	
marinara sauce	
nondiet soft drink	

 STOP!   ANSWER ALL QUESTIONS ABOVE BEFORE PROCEEDING.

2. Describe the contents of your stomach 1 hour after eating this meal. What, if anything, has happened to the chemical constituents of the meal? What, if anything, has been digested, and what, if anything, has been absorbed?

CHEMICAL CONSTITUENTS	WHAT HAPPENS TO . . . IN THE STOMACH
carbohydrates	
proteins	
fats	
vitamins	
electrolytes	
water	

STOP!   ANSWER ALL QUESTIONS ABOVE BEFORE PROCEEDING.

3. Digestion and absorption of carbohydrates, proteins, and fats are largely completed in the small intestine. Describe the process of digestion and absorption for each of these nutrients in the small intestine.

	DIGESTIVE ENZYMES	WHERE THEY COME FROM	PRODUCTS OF DIGESTION	MECHANISM OF ABSORPTION
carbohydrates				
proteins				
fats				

4. Where and by what mechanism(s) are vitamins, water, and electrolytes absorbed?

	Site of absorption	Mechanism
vitamins		
water		
electrolytes		

👉 *You have just completed Problem 5 in Unit 9.*
Please turn the page to begin work on the next problem.

10

Body Fluid Compartments

Problem 1

PROBLEM INTRODUCTION

Excessive gains or losses of water and/or electrolytes cause changes in the body fluid compartments. Such disturbances can affect the volume, composition, and concentration of body fluids, and this can have profound effects on many bodily functions.

The distribution of the body fluids of a man weighing 75 kg was being evaluated because of a suspected problem. He was injected with 100 curies (Ci), a measure of radioactivity, of 35 S-sulfate (radiosulfate) intravenously. Twenty minutes after the injection, a blood sample was drawn. It contained 0.0057 Ci of radiosulfate per milliliter of plasma water. In the 20-minute equilibration period, 4 Ci of the injected sulfate was lost in the urine. Assume that there were no other losses and that the tracer had distributed uniformly in the compartment that it entered.

1. Calculate the volume into which the sulfate distributed.

 STOP! ANSWER ALL QUESTIONS ABOVE BEFORE PROCEEDING.

The same subject was simultaneously injected with 99.8 ml of D_2O (heavy water). Two hours after the injection, a blood sample was drawn. It contained 0.232 g of D_2O per 100 ml of plasma water. In the 2-hour equilibration period, 0.4 g of the injected D_2O was lost in the urine. Assume that there were no other losses and that the D_2O had distributed uniformly in the compartment that it entered.

2. Calculate the volume into which the D_2O distributed.

STOP!   ANSWER ALL QUESTIONS ABOVE BEFORE PROCEEDING.

3. What are the subject's:

COMPARTMENT	VOLUME (L)
total body water	
intracellular fluid	
extracellular fluid	

4. Calculate the fluid volumes for a normal 75-kg man of average build, and enter these values (NORMAL) and the values that were determined for the subject (SUBJECT) in the table below. Calculate and enter the difference between the two sets of values (DIFF).

COMPARTMENT	SUBJECT (S)	NORMAL (N)	DIFF. (S - N)
total body water			
intracellular fluid			
extracellular fluid			

STOP!   ANSWER ALL QUESTIONS ABOVE BEFORE PROCEEDING.

5. What could have happened to the subject to produce these values? Explain your answer.

Problem 2

PROBLEM INTRODUCTION

Body fluid volumes are affected by any imbalance between input and output. The volume and composition of the excess intake or loss determine the effect on the size and concentration of the individual fluid volumes. This problem deals with the effects of drinking tap water.

A normal 60-kg man of average build quickly drank 2 liters of tap water. Before this, his osmolar concentration and fluid compartment sizes were normal. Calculate the size and osmolarity of his body fluid compartments before he drank. Also calculate the size and osmolarity of his compartments after he drank and absorbed the water but before he excreted any of it. Enter the results of your calculations in the table below.

TIME	TOTAL BODY WATER			EXTRACELLULAR FLUID			INTRACELLULAR FLUID		
	VOL	QUAN	CONC	VOL	QUAN	CONC	VOL	QUAN	CONC
Before									
aImmed									
bStstate									
cVol Change									

aImmed = immediate effect.   bStstate = final steady-state.   cVol Cg = final volume change.

👉 *You have just completed Problem 2 in Unit 10.*
Please turn the page to begin work on the next problem.

Problem 3

PROBLEM INTRODUCTION

Gain or loss of water and/or electrolytes affects the volume, composition, and concentration of body fluids, and changes in these fluid compartment properties can have significant effects on body functions.

A 60-kg man of average build was rapidly infused with 0.5 L of 2.925% NaCl. Assume that he excretes neither salt nor water.

1. What would be the final volume and osmolarity of his fluid compartments? Show your calculations and enter your answers in the table below.

TIME	TOTAL BODY WATER			EXTRACELLULAR FLUID			INTRACELLULAR FLUID		
	VOL	QUAN	CONC	VOL	QUAN	CONC	VOL	QUAN	CONC
Before									
aImmed									
bStstate									
cVol Change									

aImmed = immediate effect.   bStstate = final steady-state.   cVol Change = final volume change.

 STOP!   ANSWER ALL QUESTIONS ABOVE BEFORE PROCEEDING.

2. How much water would he have to drink to restore the osmolarity of his body fluids to normal? (Assume no excretion of either salt or water.)

3. What would be the volume of his fluid compartments after he absorbed this water? Enter your answers in the table below.

TIME	TOTAL BODY WATER			EXTRACELLULAR FLUID			INTRACELLULAR FLUID		
	VOL	QUAN	CONC	VOL	QUAN	CONC	VOL	QUAN	CONC
Before									
aImmed									
bStsate									
cVol Change									

aImmed = immediate effect.   bStstate = final steady-state.   cVol Cg = final volume change.

 You have just completed Problem 3 in Unit 10.
Please turn the page to begin work on the next problem.

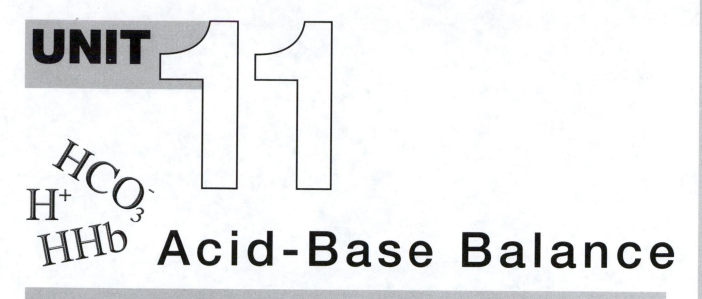

UNIT 11

Acid-Base Balance

Problem 1

PROBLEM INTRODUCTION

The status of individuals with acid-base problems may be determined from the blood levels of the components of the carbonic acid buffer system. Plotting these values on an acid-base nomogram, such as the Davenport nomogram, helps one evaluate the individual's status. It also provides information that allows one to infer aspects of the individual's history and possible causes for the condition that exists.

The acid-base status of five individuals is given below. Plot the data for each individual on the Davenport nomogram (Figure abase 1-1).

INDIVIDUAL	pH	HCO_3^- (mM)	PCO_2 (mm Hg)
A	7.35	33.0	60
B	7.25	34.1	80
C	7.28	18.3	40
D	7.50	23	30
E	7.51	36	46

Fill in the table below for the five individuals.

SUBJECT	A	B	C	D	E
Disturbance (resp/met, acid/alk [a])					
Compensation (none/part/full, resp/kidney)					
What will happen next?					
Base excess/ deficit, amount					
Possible cause					

[a] Resp/met = respiratory or metabolic; acid/alk = acidosis or alkalosis.

STOP!   ANSWER ALL QUESTIONS ABOVE BEFORE PROCEEDING.

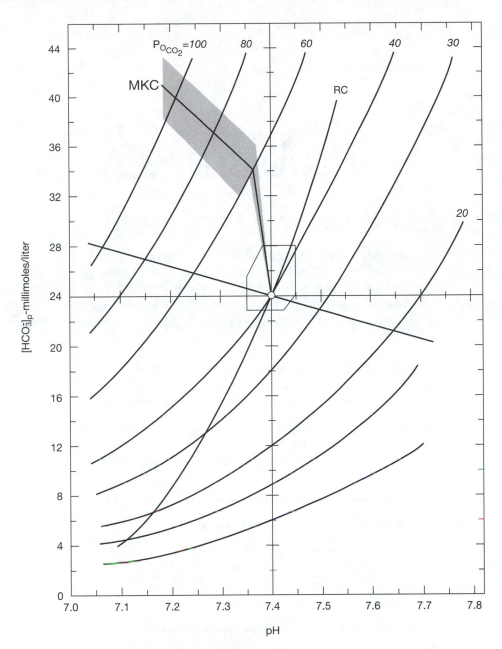

Figure abase 1-1. The Davenport nomogram is a graph of HCO_3^- versus pH that can be used to plot human acid-base data points. Points of constant P_{CO_2} are shown at the intervals (isobars). The data points of a normal individual would appear in the central polygon. An average normal individual would have the values shown as a circle within the polygon. Maximum possible respiratory compensation and maximum possible kidney compensation are shown as lines RC and MKC, respectively. The sloping line passing through the normal point is the hemoglobin titration line for a hemoglobin concentration of 15 gm%.

You have just completed Problem 1 in Unit 11.
Please turn the page to begin work on the next problem.

Problem 2

PROBLEM INTRODUCTION

Acid-base homeostasis is essential for normal function because changes in pH affect all proteins (enzymes, receptors, etc.). Hence, rapidly acting mechanisms exist that buffer any pH changes that might occur as a result of excess acid or base accumulation or loss. In addition, powerful compensatory mechanisms exist that permanently, but more slowly, correct any deviations from normal pH.

Jenny is an insulin-dependent diabetic (see Endocrine 3). She was admitted to the hospital in a coma with ketonuria because she had not taken her insulin. Ketonuria occurs when the plasma ketone level rises high enough (ketonemia) for the filtered ketone load to exceed the renal threshold. Ketones are acids.

1. Predict the changes from her normal values (increase, decrease, no change) that you would expect her to exhibit as a result of her ketonemia.

VARIABLE	CHANGE
arterial pH	
arterial $[HCO_3^-]$	
arterial PCO_2	
noncarbonic buffer base[a]	
total base[b]	

[a]Noncarbonic buffer base = total body content of all noncarbonic acid conjugate base.

[b]Total base = total base content of the body: bicarbonate + noncarbonic acid conjugate base.

 STOP! ANSWER ALL QUESTIONS ABOVE BEFORE PROCEEDING.

2. Explain the cause of these changes.

3. What is the first compensatory response that would occur (chemical buffering is not a compensatory response) to help restore her pH to normal?

4. What is the stimulus for that compensation, and on what receptor system does it act?

STOP! ANSWER ALL QUESTIONS ABOVE BEFORE PROCEEDING.

The first compensatory response to occur is respiratory compensation.

5. Predict the changes (increase, decrease, no change) that this compensation would produce from her state *after the onset of ketonemia*

VARIABLE	CHANGE
arterial pH	
arterial [HCO_3^-]	
arterial PCO_2	
noncarbonic buffer base[a]	
total base[b]	

[a]Noncarbonic buffer base = total body content of all noncarbonic acid conjugate base.

[b]Total base = total base content of the body, bicarbonate + noncarbonic acid conjugate base.

6. Explain your predictions.

STOP! ANSWER ALL QUESTIONS ABOVE BEFORE PROCEEDING.

7. Predict the changes *from normal* in her acid-base variables (increase, decrease, no change) at the completion of respiratory compensation (about 95% complete in 12 hours).

VARIABLE	CHANGE
arterial pH	
arterial $[HCO_3^-]$	
arterial PCO_2	
noncarbonic buffer base[a]	
total base[b]	

[a] Noncarbonic buffer base = total body content of all noncarbonic acid conjugate base.

[b] Total base = total base content of the body: bicarbonate + noncarbonic acid conjugate base.

STOP! ANSWER ALL QUESTIONS ABOVE BEFORE PROCEEDING.

8. Why does respiratory compensation not completely restore Jenny's arterial pH to normal?

About the time of completion of respiratory compensation, Jenny was treated with insulin.

9. What is the next acid-base compensatory process that would occur?

STOP!   ANSWER ALL QUESTIONS ABOVE BEFORE PROCEEDING.

10. Predict the changes in her acid-base variables (increase, decrease, no change) that would occur during renal compensation.

VARIABLE	CHANGE
arterial pH	
arterial [HCO$_3^-$]	
arterial PCO$_2$	
noncarbonic buffer base[a]	
total base[b]	

[a] Noncarbonic buffer base = total body content of all noncarbonic acid conjugate base.

[b] Total base = total base content of the body: bicarbonate + noncarbonic acid conjugate base.

STOP! ANSWER ALL QUESTIONS ABOVE BEFORE PROCEEDING.

11. What renal processes give rise to the increase in her plasma [HCO$_3$$^-$] during compensation?

12. Compared with normal, what is the value of arterial pH at the completion of renal compensation (increased, decreased, normal)?

13. What is the mechanism for the change in her arterial PCO$_2$ during renal compensation?

14. Plot points on the Davenport acid-base nomogram shown in Figure abase 2-1 that could reasonably represent Jenny's acid-base status:

    a. When she is normal

    b. As a result of ketonemia but before compensation

    c. At the completion of respiratory compensation

    d. Partway through renal compensation

    e. At the completion of renal compensation

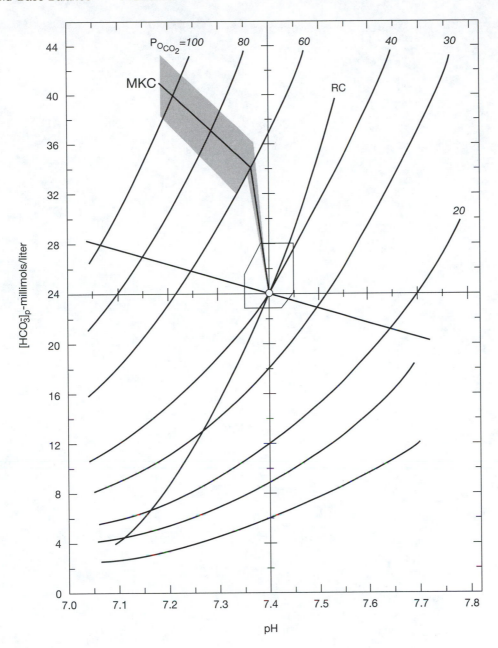

Figure abase 2-1. A Davenport acid-base nomogram. See Figure abase 1-1 for explanation.

You have just completed Problem 2 in Unit 11.
Please turn the page to begin work on the next problem.

Answers

Cell 1

OSMOLARITY/OSMOSIS

PHYSIOLOGICAL BACKGROUND

The presence of dissolved materials affects the concentration or activity of water. The size of this effect increases with the concentration of particles in the solution. Thus a molecule of NaCl, which dissociates into Na^+ and Cl^-, has the same effect on water as two molecules of glucose, which does not dissociate. The concentration of dissolved particles in a solution is its osmolarity.

When solutions of different osmolarity are separated by a water-permeable barrier, water will diffuse (osmose) from the solution of lower osmolarity to the one with higher osmolarity (i.e., the more concentrated one). Because most cell membranes are water permeable, movement of water between the intracellular fluid and the extracellular fluid will occur when the two compartments have different osmolar concentrations. Osmotic water movement can have dramatic effects on cell volume and on the concentration of materials within cells. It can thereby affect cellular function.

CLINICAL BACKGROUND

Dehydration is caused by the excessive loss of body fluids. The consequence of this loss depends on what and how much fluid is lost. In every case, there is a reduction in body fluid volume. If the lost fluid has a concentration or a composition that is different from body fluids, there will also be a change in the concentration and/or the composition of body fluids. Whenever the concentration in one body fluid compartment changes, osmosis will occur and the equality of the osmotic concentration in the fluid compartments will be restored. The equalized concentrations, however, will not be normal.

PROBLEM INTRODUCTION

Dehydration is a serious but common condition. It changes body fluid volumes, and sometimes it changes the body fluid composition and concentration. These alterations can cause a redistribution of fluid between compartments that can seriously alter function. This problem deals with the movement of water (osmosis) between red blood cells and the extracellular fluid.

Ed Rivers, a third-year medical student, was alone in the hospital emergency room (ER) one night. It was unusually quiet in the ER, so the residents were getting some much-needed sleep. A patient, Ms. M.B., was brought in showing signs of serious dehydration. Ed tried to give her water, but she vomited. Feeling that he must try something else and not wanting to wake the residents, he administered 1 liter (L) of sterile distilled water intravenously (IV).

The questions that follow are aimed at determining the consequences of Ed's action. Assume for simplicity that red blood cells (RBC) contain only solutes to which the RBC membrane is impermeable. The osmolarity of the RBC, the concentration of solute particles, is 300 mOsm/L.

1. Predict the direction of change (increase, decrease, no change) you would expect Ed's infusion to have produced in the listed parameters. Explain your predictions.

PARAMETER	PREDICTION
Ms. M.B.'s plasma osmolarity after the infusion	DECREASE
Ms. M.B.'s RBC volume after the infusion equilibrates with blood	INCREASE

Because RBC membranes are freely permeable to water, her plasma and RBC would have been in osmotic equilibrium before the infusion. The added distilled water, however, would have expanded her plasma volume and diluted her plasma solutes. Her plasma osmolarity, would, therefore, have become lower than her RBC osmolarity, and water would osmose (diffuse) from her plasma into her RBC. Hence, her RBC volume would increase while her plasma volume would fall. As water entered her RBC, the RBC osmolarity would decrease and her plasma osmolarity would increase until the osmolarity of the two solutions were again equal. After equilibration, however, the osmolarity of both would be lower than before Ed gave her the infusion.

2. The volume of Ms. M.B.'s plasma was 3 L before Ed administered the IV. Assume that there was complete mixing of the administered water with her plasma but no mixing with her interstitial fluid. Calculate Ms. M.B.'s plasma osmolarity after the infusion mixed with her plasma but before any water entered her RBC.

Her initial plasma osmolarity was 300 mOsm/L. Hence, The total number of mOsm of solute in her plasma is thus her osmolarity times her plasma volume.

    total number mOsm in her plasma = 300 mOsm/L × 3 L

    <u>total number mOsm in her plasma = 900 mOsm</u>

After infusing 1 L of water, her plasma volume becomes 4 L, but the total number mOsm in her plasma are unchanged:

        her new plasma osmolarity = 900 mOsm/4 L

        <u>her new plasma osmolarity = 225 mOsm/L</u>

This is more dilute than before the water was added.

3. Assume that Ms. M.B.'s RBC volume was 2 L before the IV was given. Calculate her plasma and RBC osmolarity after the infused water equilibrated between her plasma and her RBC. How much would her RBC volume change? (Assume that there was no mixing with her interstitial fluid.)

To determine this we need to know what the osmolarity (C) at equilibrium is. From that and our knowledge of the osmolar content of the RBC (Q), we can determine the equilibrium RBC volume (V):

$$V(L) = \frac{Q\,(mOsm)}{C\,(mOsm/L)}$$

        initial RBC volume = 2 L

       initial RBC osmolarity = 300 mOsm/L

      RBC osmoles = 300 mOsm/L × 2 L = 600 mOsm

     plasma osmoles = 900 mOsm (see Question 1)

Therefore,

$$\text{total blood mOsm} = 600 \text{ (RBC)} + 900 \text{ (plasma)} = 1500 \text{ mOsm}$$

At equilibrium,

$$\text{osmolar concentration} = \text{total mosmoles/total volume}$$

$$= 1500 \text{ mOsm/6 L}$$

Final osmolar concentration = 250 mOsm/L

At equilibrium,

$$\text{RBC volume} = \text{total RBC mOsm/concentration}$$

$$= 600 \text{ mOsm/250 mOsm/L}$$

Equilibrium RBC volume = 2.4 L.

Her RBC volume change was 2. 4L − 2 L = 0.4L increase.

Her plasma volume also increased.

$$\text{plasma volume} = \text{total volume} - \text{RBC volume} = 6 \text{ L} - 2.4 \text{ L} = 3.6 \text{ L}.$$

plasma volume increase = 0.6 L

Note that at equilibrium the 1 L of water that was infused distributed according to the distribution of osmoles. Forty% of the osmoles were in the RBC, and 40% of the added water (0.4 L) ended up there. Sixty% of the osmoles were in the plasma, and 0.6 L of the infused water ended up there.

Ms. M.B. got much worse after Ed's treatment so the resident was called. She drew some blood from the patient and centrifuged it. The denser RBC's were forced to the bottom of the tube. The supernatant plasma was pink. The resident then decided to infuse a sucrose solution into the patient (the RBC membrane is impermeable to sucrose).

4. Why was the patient's plasma pink?

At the point where the water infusion entered Ms. M.B.'s vein, the plasma would have been greatly diluted. This would have created a local situation in which the diffusion of water from the plasma into the RBCs would have proceeded rapidly, greatly expanding the RBCs and causing them to burst and release their hemoglobin into the plasma (hemolysis). This is why most infusions are iso-osmotic (have the same osmolar concentration as RBC). The pink color of the plasma results from the free hemoglobin in solution.

5. Had Ed given Ms. M.B. a sucrose solution instead of water, he might have helped her. What concentration sucrose solution should he have administered to leave her RBC volume unaffected?

To prevent water movement from or to her RBC, Ed would have had to infuse a solution that was iso-osmotic to the RBC contents, that is, 300 mOsm/L. Because sucrose is a nonelectrolyte and does not dissociate, that would be a 300 mM sucrose solution.

The resident, knowing what Ed had done, infused the patient with 1 L of a 600 mM (i.e. 600 mOsm/L) sucrose solution. Recall that, after Ed's treatment, the patient's plasma volume was 3.6 L, her RBC volume was 2.4 L, and her osmolarity was 250 mOsm/L. Assume that this infusion equilibrated with Ms. M.B.'s plasma and RBC's.

6. Predict (increase, decrease, no change) the effect of the infusion given by the resident on the parameters listed. Explain your predictions.

PARAMETER	PREDICTION
Ms. M.B.'s RBC osmolarity	INCREASE
Ms. M.B.'s RBC volume	DECREASE

The infused sucrose solution was hyperosmotic to the plasma and the RBC. Its infusion would increase the plasma osmolarity and create an osmotic gradient between the plasma and the RBC. Water would diffuse from the more dilute RBC to the more concentrated plasma, increasing the plasma volume and decreasing the RBC volume. As water left the RBC, their contents would become more concentrated while the plasma became more dilute. This process would continue until the plasma and the RBC once again had the same osmolar concentration. In the new steady state, the osmolar concentration of both would have increased.

7. Assuming that none of the administered fluid was excreted, calculate Ms. M.B.'s plasma osmolarity and RBC volume after the resident's infusion had equilibrated with her plasma and RBC.

At equilibrium, the total blood solute will be dissolved in the total volume:

1 L of 600 mOsm/L sucrose contains 600 mOsm of solute.

total solute (mOsm) = 600 (sucrose) + 900 (plasma) + 600 (RBC) = 2100 mOsm

Total volume (L) = 1 (sucrose) + 3.6 (plasma) + 2.4 (RBC) = 7 L

Final concentration = 2100 mOsm/7 L = 300 mOsm/L

The infusion has returned her concentration to normal.

RBC volume at equilibrium = 600 mOsm/300 mOsm/L = 2 L

Restoring her osmolar concentration to normal with the sucrose infusion also restored the RBC volume.

Cell 2

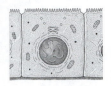

BIOELECTRICS/OSMOSIS

PHYSIOLOGICAL BACKGROUND

Intracellular and extracellular solute concentrations and the permeability of cell membranes to these solutes have important effects on cellular function. The distribution of some ions is directly determined by the activity of membrane-bound ATP-dependent pumps. For example, the concentrations of Na^+ and K^+ in the cell interior and the extracellular fluid are not the same. The concentration differences are determined by the Na^+/K^+-ATPase (also known as the Na^+ pump), which is found in the membrane of all cells.

The intracellular and extracellular concentrations of each ion determines its equilibrium potential, E_{ion}, which has two important interpretations. E_{ion} is the membrane voltage that exactly balances the tendency of the ion to diffuse down its concentration gradient. E_{ion} is also the voltage that would be created by the diffusion of the ion down its concentration gradient to a state of electrochemical equilibrium. Further, if the plasma membrane of a cell is permeable to Na^+ and K^+, they diffuse across it down their concentration gradients. In doing so, they create the resting membrane potential. The plasma membrane of most cells is very permeable to K^+. So, K^+ contributes most to determining the resting potential, which is usually close to the K^+ equilibrium potential (E_K). The activity of the Na^+ pump also directly affects the resting membrane potential, because it pumps 3 Na^+ out of the cell for each 2 K^+ it pumps into the cell (it is electrogenic).

The Nernst equation can be used to calculate the value of an ion's equilibrium potential:

$$E_{ion} = (60 \text{ mV/ion valence}) \times \log_{10}\frac{[ion]_{out}}{[ion]_{in}}$$

The distribution of some ions is not directly determined by membrane-bound pumps. The resting membrane potential determines the distribution of such passively distributed ions, and the equilibrium potential of such ions exactly equals the membrane potential.

Another aspect of cell function determined by the distribution of solute materials is the transfer of water across cell membranes. This passive process depends on the relative concentrations of dissolved particles in the extra- and intracellular fluids. When these concentrations are unequal, osmotic transfer of water through the cell membrane occurs, causing the cell volume to change until the osmotic concentrations again equalize.

PROBLEM INTRODUCTION

The intracellular and extracellular concentrations of solutes (electrolytes and nonelectrolytes) have important effects on cell function. Differences in total solute concentration in the two locations affect the distribution of water, whereas ionic distributions and permeabilities affect the resting membrane potential and the action potential.

Figure cell 2-1 shows the distribution of solute inside and outside of a hypothetical cell that is in a steady state. Assume that the cell membrane is freely permeable to water.

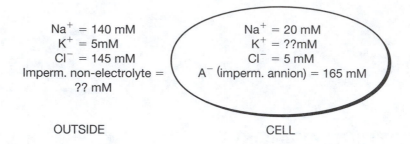

$$Na^+ = 140 \text{ mM}$$
$$K^+ = 5 \text{mM}$$
$$Cl^- = 145 \text{ mM}$$
Imperm. non-electrolyte =
?? mM

$$Na^+ = 20 \text{ mM}$$
$$K^+ = ?? \text{mM}$$
$$Cl^- = 5 \text{ mM}$$
$$A^- \text{ (imperm. annion)} = 165 \text{ mM}$$

OUTSIDE   CELL

Figure cell 2-1. The solute composition of a cell and its extracellular fluid are shown. Question marks (??) represent unknown concentrations.

1. What is the concentration of K^+ inside the cell?

   To answer this question, apply the principle of <u>electroneutrality</u>. In any compartment of the body, the sum of the positive charges "macroscopically" equals the sum of the negative charges. All cells have a <u>very small</u> net positive charge on the outside of their plasma membrane and an equal <u>very small</u> net negative charge on the inside. This charge separation is miniscule, however. To produce a potential difference of 100 mV (larger than most resting membrane potentials) in a 100-micron-diameter cell (approximately the size of a heart muscle cell) requires the transfer of an excess of approximately 1 ion per billion of the total ion content of the cell. Hence, "macroscopic" electroneutrality holds.

   The inside of the cell in this problem contains 165 mM A^- and 5 mM $Cl^- = 170$ mM of negative ions. To be electroneutral, the cell must contain 170 mM of positive ions. It has 20 mM of Na^+ and an unknown concentration of K^+. The $[K^+]_i$ must be 170 mM total − 20 mM $Na^+ = $ <u>150 mM</u>.

2. If the cell is in a steady state, what concentration of nonelectrolyte must be present outside the cell?

   The cell in this problem is typical of most cells, and its plasma membrane is freely permeable to water. Hence, no osmotic gradient can be sustained across the cell membrane because water will diffuse across the membrane whenever there is an osmolar concentration difference between the inside and the outside of the cell. Osmotic concentration is determined by the concentration of "particles" in solution. Each molecule of a nonelectrolyte and each ion of an electrolyte (e.g. Na^+ or Cl^-) constitutes a particle.

   The inside of the cell in this problem has an osmotic concentration of 340 mOsm (i.e., 340 mM of particles): 20 mM of Na^+ + 150 mM of K^+ + 5 mM of Cl^- + 165 mM of A^-. Outside the cell there are 140 mM of Na^+ + 5 mM of K^+ + 145 mM of $Cl^- = 290$ mM.

   To prevent water movement there must be 340 mM outside the cell, too: 340 mM − 290 mM = <u>50 mM of nonelectrolyte must be present outside the cell.</u>

3. Chloride ions are passively distributed across the cell membrane. What is the membrane potential?

   The intracellular and extracellular concentrations of an ion that is passively distributed across a membrane are determined by the membrane potential. Passively distributed ions have an equilibrium potential that exactly equals the membrane potential.

   The equilibrium potential for an ion can be determined from the Nernst equation:

$$E_{ion} = (60 \text{ mV/ion valence}) \times \log_{10} \frac{[ion]_{out}}{[ion]_{in}}$$

   Membrane potential = E_{Cl} (since Cl^- is passively distributed)

$$E_{Cl} = \frac{60 \text{ mV}}{-1} \times \log \frac{145}{5}$$

$$= -60 \times \log 29$$

$$= -60 \times 1.46$$

$$\underline{E_{Cl} = -87.7 \text{ mV} = \text{membrane potential}}$$

4. What is the K+ equilibrium potential? Is it the same as the membrane potential? If not, why not?

The K+ equilibrium potential (EK) may also be calculated from the Nernst equation:

$$E_K = \frac{60 \text{ mV}}{+1} \times \log \frac{[K^+]_{out}}{[K^+]_{in}}$$

$$= 60 \text{ mV} \times \log \frac{5}{150}$$

$$= 60 \text{ mV} \times -1.48$$

$$\underline{E_K = -88.8 \text{ mV}}$$

E_K (88.8 mV) is more negative than the membrane potential (−87.7 mV). If K+ were the <u>sole</u> permeable ion that was not passively distributed, the membrane potential would equal E_K. Therefore, there must be at least one permeable ion with an electrochemical force opposing that of K+. In this case, that can only be Na+, because it is the only permeable ion with a positive equilibrium potential. In fact, the inward leak of Na+ opposes the effect of the outward diffusion of K+ in most cells. Hence, cells have membrane potentials slightly to very much more positive than E_K, depending on the size of their membrane Na+ leak.

Cell 3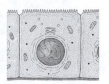

ACTION POTENTIAL CONDUCTION

PHYSIOLOGICAL BACKGROUND

When current enters the axon of a nerve cell, as occurs at a node of Ranvier during an action potential, it spreads passively down the axon. But the current density decreases with distance because part of the current is lost: some is stored in the capacitance of the membrane, and some leaks across the membrane out of the cell. The current that reaches points down the length of the axon can be increased by (1) increasing the current that enters at the action potential site, (2) decreasing the longitudinal resistance to current flow, (3) increasing the membrane resistance, and (4) decreasing the capacitative current loss. The current that enters during the action potential is determined by the E_{Na} and the number of voltage-sensitive Na^+ channels that open. Longitudinal resistance is inversely related to the axon diameter. Myelin decreases both the capacitative and transmembrane current loss.

In a myelinated axon, the voltage-sensitive Na^+ channels are concentrated at the nodes of Ranvier, while myelin greatly reduces the current loss in the internodes. This combination enables successive internodes to depolarize rapidly (saltatory conduction), confining the action potential to the nodal regions and increasing the velocity of the action potential conduction down the axon.

CLINICAL BACKGROUND

Multiple sclerosis is the most common demyelinating disease (10 to 60 per 100,000 persons in the United States). It most frequently begins between the ages of 20 and 40 and gets progressively worse over time. It involves the central white matter of the central nervous system and causes some of or all the following symptoms: weakness, abnormal reflexes, incoordination, paresthesias (tingling or numbness), and visual problems. The disease may be of viral or autoimmune origin.

PROBLEM INTRODUCTION

Many axons are covered with myelin which functions importantly in saltatory nerve conduction. There are, however, conditions in which the myelin sheath around certain axons is destroyed. This problem explores some of the possible consequences of demyelination.

A diagram of an axon with a demyelinated segment is shown in Figure cell 3-1.

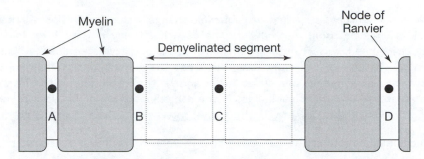

Figure cell 3-1. A segment of an axon. The shaded regions are myelinated. A and D are nodes of Ranvier between myelinated internodes. B is a node of Ranvier lying at the edge of a demyelinated region, and C is a "node of Ranvier" lying completely within the demyelinated region.

Assume that an action potential (AP) is generated at point A and propagates toward point D. Assume also that the channels in the axon membrane of the demyelinated segment are present at the same density and in the same location as in a normal axon. For each of the five parameters listed below, make the qualitative comparisons (greater, less, the same) requested. Explain your predictions. You will find this easiest to do if you first diagram the cause-and-effect relationships that are at work. RP = resting membrane potential, AP = action potential.

PARAMETER	PREDICTION
1. RP at C relative to that at A	THE SAME
2. Longitudinal current at C when the peak of an AP is at A in the de (myelinated) myelinated axon compared with the current at C in a normally myelinated axon	LESS
3. Transmembrane current at C when an AP is at B compared with the transmembrane current at B when an AP is at A	LESS
4. AP conduction velocity from A to D compared with a normal axon	LESS
5. AP amplitude at D compared with AP amplitude at A	SAME OR NONE

1. Resting potential at C relative to A

The RP is determined by the resting ion permeabilities and the extracellular and intracellular concentrations of all ions that are not passively distributed across the membrane. In axons the relevant ions are K^+ and Na^+ with the membrane being very much more permeable to K^+ than to Na^+. Hence, the RP is usually very near E_K (the K^+ equilibrium potential).

Myelin has no effect on the resting permeability to K^+ or Na^+, and the distribution of these ions is determined by the action of the Na^+/K^+-ATPase (Na pump). Hence, <u>demyelination has no effect on the resting membrane potential at C.</u>

2. Longitudinal current at C when the peak of an AP is at A in the demyelinated axon compared with the current at C in a normally myelinated axon

We have assumed that demyelination caused no change in the distribution of voltage-dependent channels along the axon, as would be the case if the myelin had been lost rapidly. Also, demyelination has no effect on the axon diameter, the primary determinant of longitudinal axon resistance. The loss of myelin, however, increases the current lost in charging the membrane capacitance and increases the current leak across the membrane in the internodal regions. Hence, an excess of current will be lost in the demyelinated region between B and C. Therefore, the longitudinal current flow at C in the demyelinated axon is less than in a normal axon.

3. Transmembrane current at C when an AP is at B compared with the transmembrane current at B when an AP is at A

The longitudinal spread of current depends on how much current enters the axon during the upstroke of the AP and how much is lost between the site of the AP and the reference site. In turn, the transmembrane current at any reference site depends on how much current reaches it.

A normal AP occurs at both points A and B. But, the segment between B and C is demyelinated, whereas the segment between A and B has myelin. Hence, more current will be lost between B and C than between A and B. So, more current will reach B when the AP is at A than will reach C when the AP is at B. Hence, more current will cross the membrane at B than at C.

4. AP conduction velocity from A to D compared with a normal axon

Myelin decreases the membrane capacitance and increases its resistance. Hence, it increases the longitudinal spread of current from the site of the action potential. When demyelination occurs, some of the current that normally flows down the axon is lost through or to the membrane.

The current loss between points A and D is increased because of the demyelination. Hence, in a demyelinated axon, it takes longer for enough current to flow into the distal regions (D) to depolarize them and open enough voltage-dependent Na^+ channels to cause an action potential. Therefore, the action potential velocity is reduced.

5. AP amplitude at D compared with AP amplitude at A

The amplitude of an action potential depends on the resting membrane potential, how many voltage-dependent Na^+ channels are available to be opened in the regenerative (rising) phase of the action potential, and the E_{Na} (depends on the Na concentration gradient).

Demyelination has no effect on the resting membrane potential (see Question 1), and we have assumed that it has no effect on the channel distribution or number. The E_{Na} depends on the intra- and extracellular $[Na^+]$, which are also unaffected by demyelination. Therefore, if enough longitudinal current crosses the demyelinated region to exceed the threshold at point D, a normal AP will be produced (the all-or-none principle). If, however, the current arriving at D is insufficient to depolarize the membrane to threshold (i.e., if the space constant, λ, is too small), no AP will occur there and the information normally transmitted by the neuron will be lost. Hence, all the signs and symptoms of demyelinating disease will be seen (see Clinical Background). See Figure cell 3-2.

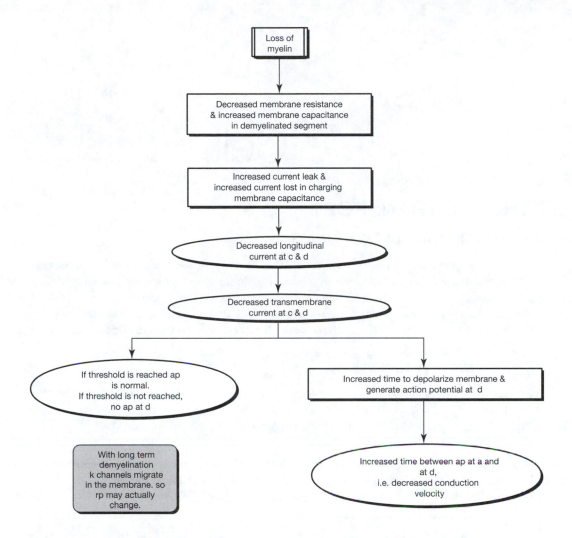

Figure cell 3-2. A causal diagram showing the effects of demyelination.

Cell 4

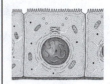

EPITHELIAL TRANSPORT

PHYSIOLOGICAL BACKGROUND

Epithelial transport results in the absorption of food, electrolytes, and water from the GI tract and the reabsorption and secretion of materials by the renal nephrons. These processes use many of the same mechanisms to accomplish their functions. One such mechanism is the cotransport of Na^+ and organic materials (glucose, amino acids). It enables the GI tract to absorb all the glucose presented to it. For this process to work effectively, however, the intestine must receive an adequate blood flow.

PROBLEM INTRODUCTION

The transport of many important solutes across epithelial cells is coupled to the transport of Na^+ and this process uses the energy content of the Na^+ concentration gradient. The intestinal epithelium is one such example.

Suppose that the blood flow to a section of the intestine was severely reduced from normal. Predict the effects (increase, decrease, no change) of this condition on the function of the enterocytes (intestinal mucosal epithelial cells) and on intestinal absorption.

PARAMETER	PREDICTION
1. Activity of the enterocyte Na^+/K^+-ATPase (Na^+ pump)	DECREASED
2. Lumen to enterocyte Na^+ concentration gradient	DECREASED
3. Glucose absorption from the intestinal lumen	DECREASED
4. Water absorption from the intestinal lumen	DECREASED

Explain your predictions here. You will find it helpful to diagram the causal relationships that are involved.

1. Activity of the enterocyte Na^+/K^+-ATPase (Na^+ pump)

   Tissue blood flow provides the nutrients and oxygen needed for oxidative phosphorylation, that is, for the normal metabolic production of ATP. With severely reduced blood flow, ATP production will be reduced. The activity of the Na^+/K^+ pump requires, and is directly coupled to, the hydrolysis of ATP. Hence, a reduction in blood flow will cause the activity of the Na^+/K^+ pump to be reduced.

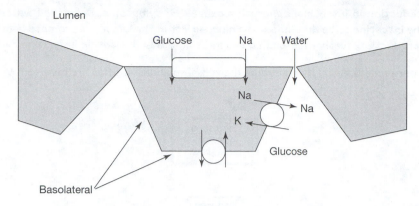

Figure cell 4-1. Three enterocytes are pictured, which separate the GI lumen from the extracellular fluid. The glucose + Na$^+$ cotransporter (symport) is located on the lumenal surface, and the Na$^+$/K$^+$-ATPase is found on the basolateral membrane. Absorption of Na$^+$ and glucose (along with other solutes) increments the osmolarity of the fluid in the lateral spaces (between adjacent enterocytes), providing the "driving force" for water absorption.

2. Lumen to enterocyte Na$^+$ concentration gradient

   Reducing the activity of the Na$^+$/K$^+$ pump will decrease the extrusion of Na$^+$ from the enterocytes and will allow intracellular [Na$^+$] to increase. Hence, the difference between the extracellular [Na$^+$] and the intracellular [Na$^+$]—the Na$^+$ concentration gradient —will be smaller.

3. Glucose absorption from the intestinal lumen

   One of the main mechanisms for the absorption of glucose from the intestinal lumen is via a secondary active transport mechanism that uses a Na$^+$/glucose symport (both molecules are transported in the same direction) protein imbedded in the lumenal membrane of the enterocytes. This transport system uses the energy present in the [Na$^+$] gradient to transport glucose against its concentration gradient, that is, Na$^+$ is transported from the lumen, where its concentration is high, into the enterocyte, where its concentration is low. At the same time, glucose is transported from the intestinal lumen, where its concentration is low, into the enterocyte, where its concentration is high. (See Figure cell 4-1.)

   Restricting the intestinal blood flow reduces Na$^+$ pumping, and the Na$^+$ concentration gradient gets smaller. Therefore, less glucose is transported via the Na$^+$/glucose symport. Normally, all the dietary glucose is absorbed. In this situation, however, some may remain behind and be excreted in the feces.

4. Water absorption from the intestinal lumen

   The absorption of water from the intestinal lumen is a passive process. It results from the active and secondary active absorption of solutes. These solutes are absorbed into the lateral spaces between adjacent epithelial cells and raise the osmotic pressure there, creating an osmotic pressure gradient that causes water movement from the lumen through the inter-epithelial tight junctions (see Figure cell 4-1).

   With the reduced absorption of Na$^+$, glucose, and other solutes cotransported with Na$^+$, the small osmotic gradient that normally drives water absorption from the lumen will decrease. Water absorption will be smaller.

The reduction in intestinal absorption causes GI symptoms. Decreased water absorption causes the intestine to be distended. The glucose left in the lumen will promote bacterial growth and gas production, further distending the intestine. This distention impairs motility.

5. Explain your predictions by diagramming the causal relationships that are involved.

See Figure cell 4-2.

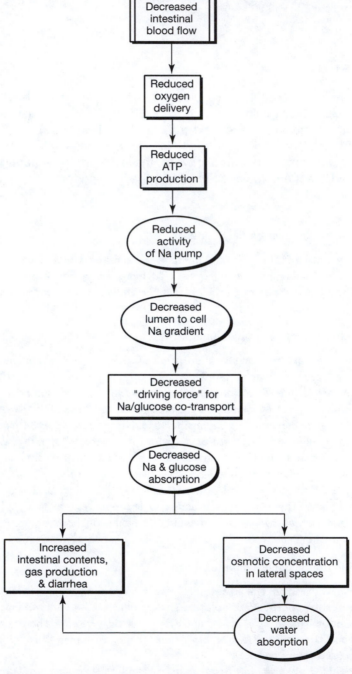

Figure cell 4-2. The causal relationships between a decrease in intestinal blood flow and a decrease in glucose, Na+, and water absorption are shown.

Muscle 1

MUSCLE WEAKNESS/MYASTHENIA GRAVIS

PHYSIOLOGICAL BACKGROUND

Voluntary motor activity is initiated in the cerebral cortex and, in sequential order, involves neural components between there and the synaptic terminals of the α-motor neurons. The axon terminals of α-motor neurons release acetylcholine (ACh), the neuromuscular transmitter, which triggers the subsequent events on and in the skeletal muscle to bring about muscle shortening and force generation.

ACh acts by binding to receptors in the motor end plate, causing nonspecific cation channels to open. This generates a local electrical event, a depolarization called the end plate potential (EPP). Normally, the EPP is large enough to depolarize the adjacent, excitable muscle membrane to its threshold, giving rise to an action potential (AP). The AP then propagates over the entire sarcolemma and penetrates into the T-tubules, providing the stimulus for excitation-contraction coupling.

During repetitive stimulation, the quantity of ACh that is released by successive motor nerve APs declines. Normally, however, even during repetitive activation, the quantity of ACh released is sufficient to produce EPPs large enough to trigger muscle APs.

CLINICAL BACKGROUND

Myasthenia gravis is the best understood of the autoimmune diseases. It is caused by an antibody reaction to one of the components of the ACh receptor, leading to a reduction in the number of functioning receptors. As a result, myasthenics have muscle weakness. That weakness may be confined to a small group of muscles (e.g, the extraocular muscles), or it may be more widespread. If the respiratory muscles become involved, the patient faces a crisis and will require ventilatory assistance. There are about 25,000 myasthenics in the United States.

The symptoms of myasthenia are muscle weakness and increased fatigability. The muscle weakness is, to some extent, present at all times, but it becomes exaggerated with repetitive activity. The weakness is a consequence of the reduced number of ACh receptors, which decreases the EPP in some of the affected muscle fibers below the value needed to trigger a muscle AP. Such fibers do not contract and do not contribute to the force generated by the whole muscle.

Fatigability results from the decline in the amount of ACh released by repetitive action potentials, a normal process. In myasthenics, this reduced quantity of transmitter has fewer receptors to react with and in some of the fibers is unable to cause an EPP of sufficient magnitude to generate a muscle AP.

The traditional treatment for myasthenia gravis is administration of anticholinesterase agents. These drugs inhibit the breakdown of ACh by acetylcholinesterase, thus increasing the amount of transmitter available to react with the reduced number of receptors. This therapy also offsets the consequences of falling transmitter release with repetitive stimulation.

ACh receptors are constantly being destroyed, and new ones are constantly being synthesized. This normal process is accelerated in myasthenics. Newer immunotherapy takes advantage of this situation. The treatment inactivates the antibodies that destroy the receptors, followed by accelerated replacement of the defective receptors with normal ones, thus "curing" the disease.

PROBLEM INTRODUCTION

The motor pathway consists of a large number of elements organized in a functional sequence. A motor action involves these elements, one after the other, to bring about shortening and force generation by skeletal muscle. Malfunction of any one of these elements can affect muscle strength and coordination.

A patient came to his physician complaining of muscle weakness.

1. List the loci along the complete motor pathway where some defect could result in muscle weakness.

LOCI IN MOTOR PATHWAY
MOTOR (CEREBRAL) CORTEX
DESCENDING SPINAL CORD AXONS
SPINAL CORD α-MOTOR NEURONS
AXON TERMINALS OF α-MOTOR NEURONS
MOTOR END PLATE
SARCOLEMMA
SARCOPLASMIC RETICULUM
MYOSIN CROSSBRIDGES
FORCE GENERATING MECHANISM

2. Upon testing the patient, the following observations were made:

   a. Repetitive voluntary activity was associated with normal action potential activity in the motor nerve, although the patient's muscles rapidly fatigue and muscle force declined.

   b. Repetitive direct electrical stimulation of muscle resulted in normally sustained force without evidence of fatigue.

Assuming that a single site of pathology is present, where along the motor pathway is this patient's problem? Explain your reasoning.

Because voluntary motor effort elicited normal action potential activity in the motor nerve, the connection between the cerebral cortex and the α-motor neurons must be functioning. Stimulating the muscle directly yielded normal muscular function. Hence, all the elements from the sarcolemma to the force generating mechanism must also be intact. What remains unevaluated are those structures from the α-motor nerve endings through the motor end plate. The patient's difficulty must be there.

Possible defects could be inadequate transmitter (acetylcholine, ACh) synthesis or release, a defect in the ACh receptors (abnormal structure or number of receptors), or a defect in the transducer function of the receptor (the opening of the cation channel). The data do not permit us to distinguish between these alternatives.

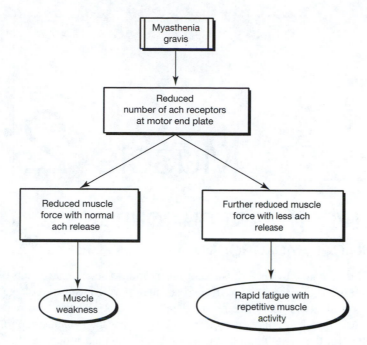

Figure musc 1-1. A concept map explaining the signs and symptoms exhibited by a myasthenic patient.

This patient has an autoimmune disease called myasthenia gravis, a condition in which some of the patient's ACh receptors have been destroyed in an antigen-antibody reaction. As a consequence, some of the patient's muscle fibers are not being activated, which is causing muscle weakness. More of the patient's muscle fibers fail to be activated during attempts at repetitive muscle activity because the amount of ACh released decreases with repetitive activity.

3. What might be done to help such patients?

There is no way to replace the ACh receptors that have been lost in the immune reaction directly. One can, however, compensate for the loss of receptors, especially in repetitive activation, when the transmitter release declines, by inhibiting the breakdown of transmitter. Acetylcholinesterase inhibitors are used for this purpose.

Newer therapy involves neutralizing the antibodies to the receptor with antigenic materials produced in other biospecies, which allows newly synthesized receptors to replace the lost ones.

Muscle **2**

EXCITATION/CONTRACTION COUPLING

PHYSIOLOGICAL BACKGROUND

Skeletal muscle is activated by way of its motor innervation. The axon terminals of α-motor neuron depolarize when an action potential (AP) arrives there, which causes voltage-dependent Ca^{2+} channels in the terminals to open. The resulting influx of Ca^{2+} leads to the release of transmitter (acetylcholine, ACh), which diffuses the short distance across the synaptic gap to the acetylcholine receptors (AChR) on the motor end plate where it binds, opening a nonspecific cation channel. A nonpropagating end plate potential (EPP) results. The local electrical current caused by the EPP depolarizes adjacent excitable regions of the sarcolemma, giving rise to a muscle AP that propagates over the sarcolemma, invades the T-tubules, and triggers the intracellular release of Ca^{2+} from the sarcoplasmic reticulum (SR). The free Ca^{2+} reacts with the troponin of the thin filament, causing a conformational change in tropomyosin and exposing the active sites on the actin filaments. Activated myosin head groups (cross bridges) react with the active sites of actin. Using the energy content of ATP, they cause muscle shortening and/or tension development. The events described are sequential and causal, as illustrated in Figure musc 2-1.

Caffeine acts by sensitizing the release of Ca^{2+} from the sarcoplasmic reticulum. It does so by making it easier for the Ca^{2+} channels to open and more difficult for them to close. Hence, at low doses it causes the contraction induced by motor nerve stimulation to start sooner, last longer, and have a higher amplitude. Larger concentrations of caffeine produce contracture without nerve stimulation by increasing the permeability of the SR for Ca^{2+}, releasing the stored Ca^{2+}, and leading to a sustained increase in intracellular Ca^{2+} concentration. This results in a maintained interaction of actin and myosin, that is, contracture.

PROBLEM INTRODUCTION

Neural activation of skeletal muscle involves a sequence of steps. Understanding the steps in this sequence helps to determine the locus of disturbances that affect muscle activation.

You are given an isolated muscle with its motor nerve intact. This preparation is arranged so that you can make electrophysiological measurements from the nerve and the muscle as well as mechanical measurements of the muscle contraction. You observe that if caffeine is added to the fluid bathing the preparation, the peak tension developed in a muscle twitch produced by stimulating the motor nerve is increased.

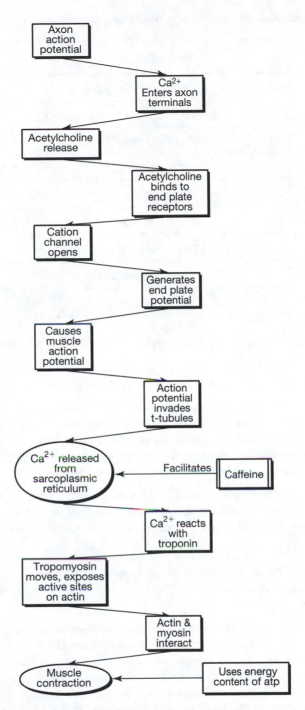

Figure musc 2-1. The sequence of steps involved in the activation of skeletal muscle by a motor nerve. Caffeine influences the release of Ca2+ from the SR at the step indicated.

1. List the *possible* loci at which caffeine may be exerting its effect.

POSSIBLE LOCI
MOTOR NERVE AXON
MOTOR NERVE ENDING Ca^{2+} CHANNELS
SYNAPTIC VESICLES
ACH RECEPTOR
SARCOLEMMA
T-TUBULE
SARCOPLASMIC RETICULUM
TROPONIN/TROPOMYOSIN
ACTIN/MYOSIN

2. Caffeine is known to alter transmembrane movement of Ca^{2+}. What is(are) the possible site(s) at which caffeine is acting?

There are two important sites where Ca^{2+} is involved.

1. The motor nerve AP opens a voltage-dependent Ca^{2+} channel in the presynaptic motor nerve endings. This action allows an influx of Ca^{2+} down its electrochemical gradient, causing the fusion of transmitter-containing vesicles with the axon terminal membrane and the release of ACh. One mechanism for the action of caffeine could be to cause increased Ca^{2+} entry into the nerve ending, producing increased transmitter release.

2. Arrival of the AP at the ends of the T-tubules triggers the release of stored Ca^{2+} from the SR. A different mechanism for the action of caffeine could be to facilitate the release of Ca^{2+} from the SR directly, which would lead to muscle contraction without the need for an action potential.

3. High concentration of caffeine causes an isolated muscle to contract in the absence of any activity in the motor nerve. This contraction is not blocked by nicotinic blocking agents like curare. Where do you think caffeine acts?

If caffeine acted on the axon endings of the α-motor neurons to cause or accentuate the release of ACh, its effect on muscle contraction would result from the action of acetylcholine. Curare blocks the effects of ACh at the motor end plate but does not block the effects of caffeine. Hence, the effects of caffeine cannot be due to the release of ACh from nerve endings.

Caffeine directly released Ca^{2+} from the SR and/or inhibited its uptake into the SR would cause contraction without the need for nerve impulses. This action would not be affected by blocking the reaction of ACh with its receptor because it occurs later in the series of activating steps than the reaction of ACh with its receptor. Hence, the action of caffeine must be on the SR.

See Figure musc 2-1 for the steps in muscle activation and the location of the site of caffeine action.

Muscle 3

MALIGNANT HYPERTHERMIA

PHYSIOLOGICAL BACKGROUND

An essential step leading from excitation of skeletal muscle to contraction is the liberation of stored biochemical energy (from the high-energy phosphate of ATP) by the myosin ATPase (located on the crossbridge). It then folllows that the muscle must have a store of ATP and the ability to generate new, replacement ATP (whether by the normal process of oxidative phosphorylation or by anaerobic pathways).

Most of the energy liberated from ATP, however, appears as heat and is not converted into mechanical energy. Skeletal muscle therefore plays an important role in temperature homeostasis. For example, muscle is used to increase heat production by shivering when body temperature must be maintained in a cold environment. On the other hand, in a 70 kg individual, skeletal muscle accounts for approximately 30 kg of body weight, and the heat generated by active skeletal muscle represents a major part of the heat load that must be dissipated by the body.

CLINICAL BACKGROUND

Malignant hyperthermia is a relatively rare response to certain forms of anesthesia. It is caused by an unknown mechanism that gives rise to an increase in Ca^{2+} release from the sarcoplasmic reticulum, producing skeletal muscle contractures (continuous, sustained contractions resulting in muscle rigidity) accompanied by a hypermetabolic state: muscle tissue oxygen use is increased and thus carbon dioxide production is increased; there is acidosis and hyperpnea (from the elevated carbon dioxide). Most importantly, there is a very rapid rise in heat production and body temperature, which can be life threatening.

This condition appears to involve an inheritable defect in muscle metabolism that may be activated by halothane anesthesia; Ca^{2+} stored in the sarcoplasmic reticulum is released into the muscle cytoplasm where it causes sustained muscle contraction. This reaction does not necessarily occur on the first exposure to this anesthetic agent, and its occurrence is not easily predicted.

Treatment for this response calls for immediate cessation of the anesthesia, cooling the body with ice packs, and the use of dantrolene, a drug that specifically reduces the release of Ca^{2+} from the SR.

PROBLEM INTRODUCTION

Activation of skeletal muscle occurs through a complex sequence of events, leading to contraction (shortening) or force generation. Abnormal responses at any of these steps will alter muscle function and possibly whole body homeostasis.

Muscle is a biochemical machine that converts chemical bond energy into mechanical force or shortening.

1. Diagram the sequence of steps from the release of transmitter from the motor nerve to the contraction of the skeletal muscle cell that occurs.

   Physiologically, skeletal muscle contracts only in response to activation of its motor nerve. Action potentials in motor nerves release acetylcholine from the nerve terminals. The transmitter diffuses across the synaptic gap and binds to receptors at the end plate, opening a cation selective channel and resulting in an end plate potential (EPP). The EPP generates an action potential in the sarcolemma that propagates over the entire muscle cell. At the T-tubules (invaginations of the sarcolemma), the depolarization of the membrane is passively conducted into the interior of the cell, causing the sarcoplasmic reticulum to release the calcium stored there. This action causes a sudden rise in cytoplasmic calcium, removing the inhibition produced by the troponin-tropomyosin complex. Myosin crossbridges are then able to react with actin. Myosin ATP-ase at the crossbridge breaks down ATP, releasing energy to be used either to shorten the muscle or to generate force.
   Figure musc 3-1 diagrams this sequence of events.

2. Biochemical processes give rise to waste products that in many cases pose a homeostatic challenge for the organism. What are the inevitable by-products of muscle contraction with which the body must cope?

   The conversion of bond energy in ATP to mechanical energy is not perfect, and heat is also generated as a waste product of this reaction. This heat, of course, poses a challenge for the body's thermoregulatory system.
   The metabolic processes in muscle generating ATP use oxygen and result in the production of carbon dioxide, another waste product that the respiratory system removes from the body.
   If the delivery of oxygen to contracting muscle cannot sustain the muscle's oxidative metabolism at the muscle's ongoing work level, the muscle begins to use anaerobic pathways to produce ATP, with the resulting appearance of lactic acid. This change represents a challenge to the body's acid-base regulating system.

A 25-year-old man was brought to the emergency room with a broken leg. His medical history was unremarkable, although he reveals that his mother had died of some "complication" following a hysterectomy.

The patient was taken to surgery for repair of his fractured left distal tibia. He was premedicated with, among other drugs, succinylcholine (a muscle relaxant) and was placed on a ventilator. It was noted that vigorous muscle fasciculation began almost immediately. Nevertheless, general anesthesia was begun with halothane.

3. Succinylcholine binds to the ACh receptors at the motor end plate. Explain the mechanism by which it could produce muscle relaxation.

   Succinylcholine binds to the acetylcholine receptors where it can act in two quite different ways. When the receptor binds a ligand (succinylcholine), the channel is opened and current flows across the end plate membrane. This process results in a maintained depolarization of the adjacent sarcolemma, inactivating the voltage-dependent Na-channels and effectively preventing this membrane from generating action potentials. Hence, muscle contract ceases (relaxation occurs). Succinylcholine can also act as a competitive antagonist of the physiological transmitter acetylcholine, thus blocking transmission at the motor end plate. Muscle contraction must then cease.

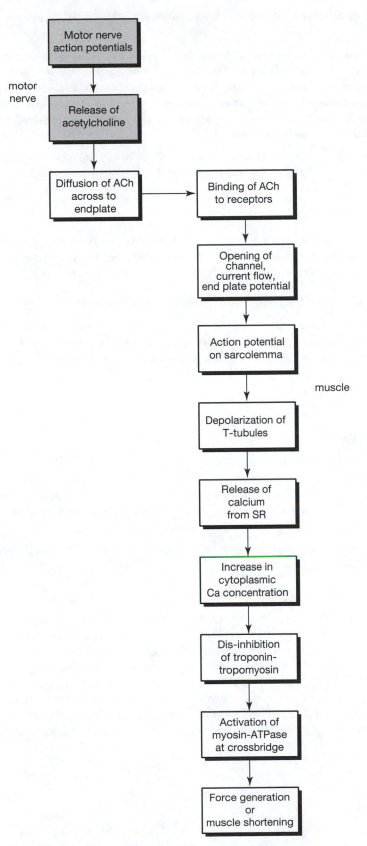

Figure musc 3-1 The sequence of steps between motor nerve activity and the contraction of a skeletal muscle.

4. Explain the mechanism by which fasciculations could occur following administration of succinylcholine.

> The initial binding of succinylcholine to the postsynaptic membrane causes the receptor channel to open, allowing current to flow and the muscle to depolarize. A brief period of repetitive firing of muscle action potentials occurs. This process activates a contraction of the fiber, but because these contractions are not coordinated, the muscle can be seen to fasciculate (quiver). Fasciculation is then followed by relaxation (flaccid paralysis) as transmission at the end plate is blocked (see the answer to Question 3).

There was an immediate tachycardia, and it was noted by the surgeon that the patient had become "as stiff as a board." The anesthetist noted that the patient's core temperature was 39.6° C and seemed to be rising rapidly.

5. What caused the patient to become "stiff as a board"?

> This patient has experienced a massive release of calcium from the sarcoplasmic reticulum as a result of an idiopathic response to halothane anesthesia. The release of calcium from the SR caused a maximal sustained muscular contraction, resulting in rigidity.

6. What could account for the patient's rapid increase in body temperature?

> Body temperature rises rapidly because the metabolic activity in skeletal muscle (30–40% of body weight) is suddenly greatly increased. Heat is a waste product of this metabolic activity, and it is being generated faster than the body can eliminate it. Body temperature must rise.

The halothane anesthesia was terminated, and the patient was cooled with ice. Dantrolene was administered intravenously. Oxygen was given.

7. Dantrolene inhibits the release of calcium from the sarcoplasmic reticulum. Explain how this would reverse the patient's hyperthermia.

> Dantrolene specifically inhibits the calcium release channels in the SR membrane, but it does not affect the SR's calcium pump. Thus calcium in the cytoplasm can be pumped back into the SR. As cytoplasmic calcium falls active sites on actin are then covered by the troponin-tropomyosin complex, actin and myosin dissociate, ATPase activity (crossbridge activity) is reduced, ATP breakdown ceases, and heat generation is reduced.

Laboratory results indicated the presence of respiratory and metabolic acidosis, hyperkalemia, and the elevation of blood lactate and pyruvate.

8. Explain these laboratory results.

> The increased muscle metabolism increases the rate of production of CO_2. Normally, respiratory control would increase ventilation so that the individual would "blow off" CO_2 as fast as it is produced. The patient, however, is being artificially ventilated (required because succinylcholine paralyzes the respiratory muscles), and the ventilatory rate does not correspond to his new, pathologic rate of CO_2 production. The accumulation of CO_2 produces a respiratory acidosis.
>
> Similarly, the increased metabolism of the muscle exceeds the rate at which O_2 is being delivered, and the muscle's metabolism becomes anaerobic. The results of this are production of lactic and pyruvic acid and a metabolic acidosis, and elevated lactate and pyruvate in the blood.

Heart Muscle 1

INOTROPIC STATE (CONTRACTILITY)

PHYSIOLOGICAL BACKGROUND

Cardiac and skeletal muscle each have more than one mechanism for changing contractile force in vivo. A mechanism that both muscle types use for changing contractile force is alteration of the preload (i.e., the muscle fiber length at the time of activation). One mechanism that changes cardiac muscle contractile force but does not affect skeletal muscle is varying the inotropic state of the muscle. This change is accomplished by changing the amount of free Ca^{2+} in the myoplasm during excitation-contraction coupling (EC coupling).

In skeletal muscle, so much activating Ca^{2+} is released from the sarcoplasmic reticulum (SR) during EC coupling that the muscle's troponin C is completely saturated, exposing all the active sites on actin. Hence, all the crossbridges in the actin/myosin overlap region can interact. The myofibrils are maximally activated. In cardiac muscle, however, the amount of Ca^{2+} in the myoplasm during EC coupling is sufficient to react with only *part* of the troponin C. Hence, cardiac contraction is usually submaximal, and cardiac muscle contraction can be incremented or decremented by varying the myoplasmic free $[Ca^{2+}]$ during EC coupling. Increasing the available Ca^{2+} increases contractility, a positive inotropic effect. Decreasing the myoplasmic Ca^{2+} has a negative inotropic effect.

Changing the inotropic state of the heart muscle affects other aspects of contraction as well, including the phosphorylation of muscle proteins.

CLINICAL BACKGROUND

Digitalis is an extract of the foxglove plant. It has been used as a drug for at least 700 years, and probably longer. Its efficacy in cardiac failure has been known since William Withering's account in the late eighteenth century. It is a member of a large class of plant-produced agents (cardiac glycosides) that are a chemical combination of a cardioactive, steroidlike moiety (aglycone) with one or more sugars. The aglycone confers the heart stimulating activity on the molecule.

PROBLEM INTRODUCTION

Cardiac muscle contractile strength can be changed by altering its inotropic state (IS). An increase in IS increases contractile activity without requiring a change in muscle fiber length. This mechanism is absent in skeletal muscle.

1. Predict the effects (increase, decrease, no change) of a positive inotropic agent (increases IS) on the following aspects of cardiac muscle contraction.

PARAMETER	PREDICTION
a. rate of isometric force development	INCREASE
b. rate of isotonic shortening	INCREASE
c. peak isometric force	INCREASE
d. peak shortening	INCREASE
e. duration of contraction	DECREASE
f. rate of relaxation	INCREASE

2. Describe the mechanism(s) that cause(s) the inotropic changes in the following contractile properties.

a. rate of isometric force development

b. rate of isotonic shortening

The rates of both isometric force development and isotonic shortening are increased by positive inotropic agents (Figure htmus 1-1, a versus b) as a result of more rapid crossbridge cycling, the probable cause for which is phosphorylation of myosin head groups. These groups are phosphorylated by protein kinases that are activated by positive inotropic agents.

c. peak isometric force

d. peak shortening

Peak isometric force and peak isotonic shortening are increased by a positive inotropic agent (Figure htmus 1-1, c versus d) without an increase in the end diastolic fiber length. The force and shortening are determined by the number of myosin crossbridges that react with active sites on actin. This number, in turn, is determined by how much Ca^{2+} enters the myoplasm from the extracellular fluid via the Ca^{2+} channels and is released from the sarcoplasmic reticulum (SR) and other intracellular Ca^{2+} stores. β-adrenergic agonists (norepinephrine, epinephrine) are the most important positive inotropic agents in vivo. They activate adenylyl cyclase, increasing cytosolic cyclic AMP concentration. This process activates (phosphorylates) protein kinases that in turn phosphorylate a control protein of the sarcolemmal Ca^{2+} channel and the SR Ca^{2+} pump. As a result, more Ca^{2+} enters the cell via the Ca^{2+} channels, and the SR accumulates (pumps) additional Ca^{2+}. The net result is an increase in cytoplasmic Ca^{2+} concentration during subsequent EC coupling reactions, an increase in Ca^{2+}/troponin reaction, an increase in the number of active sites exposed on actin, an increase in actin/myosin crossbridge formation, and an increase in the peak force or peak shortening.

e. duration of contraction

f. rate of relaxation

The duration of contraction is reduced by positive inotropic agents, and the rate of relaxation is increased (Figure htmus 1-1, e versus f and h versus g). The same phosphorylation that causes the SR to accumulate more Ca^{2+} increases the rate of uptake of Ca^{2+} from the cytoplasm to the SR. Therefore, Ca^{2+} dissociates from troponin sooner, causing the active sites on actin to be covered by tropomyosin sooner and contraction to be terminated sooner. Besides Ca^{2+} is sequestered in the SR faster, relaxation is more rapid. Also, troponin I (another muscle protein) is phosphorylated by the activated kinases. In its phosphorylated state, troponin I inhibits Ca^{2+} binding by troponin C, further facilitating relaxation.

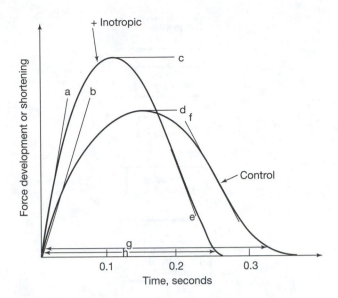

Figure htmus 1-1. Single contractions of an untreated (control) cardiac muscle and a cardiac muscle treated with a positive inotropic agent. The graphs show the force developed or the amount of shortening (ordinate) versus time. The lettered lines are drawn to allow individual features of the contractions to be compared.

3. Digitalis (ouabain) is a drug used to counteract the effects of cardiac failure. In heart failure, one or both ventricles are incapable of developing a normal contractile force or of producing a normal stroke volume without an increase in end diastolic filling. In such situations, digitalis increases the inotropic state (contractility) of heart muscle. Digitalis acts by inhibiting the Na^+/K^+ pump.

a. What effect will treatment with digitalis have on the distribution of ions across the membrane of cardiac muscle cells?

The Na^+/K^+ pump (also known as the Na+ pump and as the Na^+/K^+-ATPase) maintains the normal distribution of Na^+ and K^+. The concentration of Na^+ is low within cells and high extracellularly, whereas the concentration of K^+ is high within cells and low extracellularly. Slowing the Na^+/K^+ pump decreases the extrusion of Na^+ from cells, causing intracellular Na^+ concentration to increase and the transsarcolemmal Na^+ gradient to decrease.

b. The cardiac muscle membrane contains a secondary active transporter, a Na^+/Ca^{2+} exchanger. This countertransporter is one of the mechanisms that maintains a low cytosolic Ca^{2+} concentration during diastole. It uses the energy content in the trans-sarcolemmal Na^+ electrochemical gradient to transport 3 Na^+ into the cell for each Ca^{2+} it extrudes from the cell. Explain how the action of digitalis interacts with the action of the transporter to increase the force of contraction of the myocardium.

The Na^+/Ca^{2+} exchanger is driven by the energy present in the transcellular Na^+ gradient. This gradient is reduced by digitalis, slowing the rate of extrusion of Na^+ from the cells and allowing intracellular Na^+ concentration to increase. Hence, less Na^+ enters the cells via the Na^+/Ca^{2+} exchanger and less Ca^{2+} leaves the cells; intracellular Ca^{2+} concentration increases. In cardiac muscle, the intracellular Ca^{2+} concentration is not normally large enough to saturate the troponin C during EC coupling. Therefore, some of the myosin-reacting sites on actin are unavailable for participation in the contraction process. A digitalis-induced increase in intracellular Ca^{2+} concentration uncovers more of these sites, increasing the number available for crossbridge formation, which has a positive inotropic effect on the myocardium (contractilityis increased). Thus the force of contraction of the myocardium at any degree of filling will be increased. The sequence of events is diagrammed in Figure htmus 1-2.

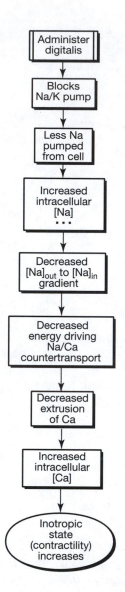

Figure htmus 1-2. A concept map showing the causal relationships that act to produce digitalis' positive inotropic effect.

The following studies were carried out on two identical isolated rabbit hearts. The ventricles were made to contract by electrical stimulation of the ventricular conducting systems. The coronary arteries of heart 1 were perfused with normal, untreated rabbit blood, whereas those of heart 2 were perfused with blood containing a positive inotropic agent (high IS). In the descriptions below, subscript "1" refers to heart 1 and subscript "2" refers to heart 2.

Two experiments were carried out. In the **first experiment**, the ventricles of heart 1 and heart 2 were filled to the same low end diastolic volume (low EDV), and stroke volumes (SV_1 and SV_2) were measured. See the conditions 1 and 3 in the table on the following page.

In the **second experiment**, the ventricles of both hearts were filled to an equal but greater EDV (high EDV) than in Experiment 1. Again the stroke volumes were measured (SV_1 and SV_2). See conditions 2 and 4 in the table on the following page.

	EXPERIMENT 1	EXPERIMENT 2
	low EDV	high EDV
normal IS (heart 1)	1	2
high IS (heart 2)	3	4

4. Predict the following (greater, smaller, same):

PARAMETER	PREDICTION
a. the size of SV_1 compared with SV_2 in Experiment 1 (1 versus 3 in the table)	SMALLER
b. the size of SV_1 with low EDV compared with SV_1 with high EDV (1 versus 2 in the table)	SMALLER
c. the size of SV_2 with low EDV compared with SV_2 with high EDV (3 versus 4 in the table)	SMALLER
d. the change (Δ) in SV_1 when going from low EDV to high EDV (Δ 1 to 2 in the table) compared with the change (Δ) in SV_2 when going from low EDV to high EDV (Δ 3 to 4 in the table)	SMALLER

5. What is the cause of the difference you predicted in Question 4, part a?

The two hearts are identical except that heart 2 is being perfused with a positive inotropic agent, which increases the cytosolic Ca^{2+} concentration during EC coupling. Therefore, at the same EDV, heart 1 which has normal IS (contractility) will contract less forcefully than heart 2, which has an increased IS. Hence, SV_1 is smaller than SV_2. See Figure htmus 1-3, 1 versus 3.

6. What is the cause of the difference you predicted in question 4, part b and c?

In each case, the EDV has been increased, thus increasing the fiber length. By Starling's law of the heart, increasing the fiber length increases the force of contraction by increasing the sensitivity of troponin to Ca^{2+} and by reducing the actin to actin interference in the midsarcomere. Thus the SV at low EDV is smaller than the SV at high EDV. See Figure htmus 1-3 1 versus 2 and 3 versus 4.

7. What is the cause of the difference you predicted in question 4, part d?

The curve relating EDV (fiber length) to ventricular performance (SV) has a smaller slope in the untreated heart (1) than in the heart with augmented contractility (2, treated with a positive inotropic agent). Therefore, increasing EDV increases SV less in heart 1 than in heart 2. See Figure htmus 1-3, Δ 1 to 3 versus Δ 2 to 4.

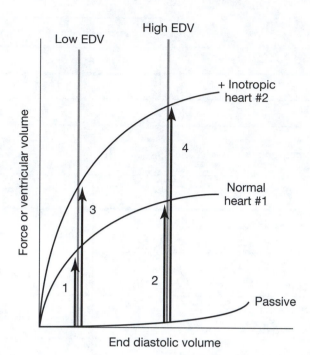

Figure htmus 1-3. A graph of heart force or ventricular volume (ordinate) versus ventricular end diastolic volume for two rabbit hearts. The lower curve is for a heart (1) perfused with untreated rabbit blood (normal), and the upper curve is for a heart (2) perfused with blood containing a positive inotropic agent. Stroke volume (SV) is the vertical distance from the passive curve to the line representing either the normal heart (SV$_1$) or the heart treated with an inotropic agent (SV$_2$). Hence, arrows 1 and 3 are the stroke volumes at low EDV. Arrows 2 and 4 are the stroke volumes at high EDV. See the text for a description of the experiment.

Heart Muscle 2

INOTROPIC STATE: CALCIUM

PHYSIOLOGICAL BACKGROUND

Unlike skeletal muscle, heart muscle depends in part on an influx of extracellular Ca^{2+} for excitation-contraction (EC) coupling. Ca^{2+} enters myocardial cells during the plateau phase (phase 2) of the action potential (the slow inward current). This Ca^{2+} has two important functions: (1) it causes Ca^{2+} to be released from the sarcoplasmic reticulum (SR), called calcium-induced calcium release, and (2) it adds to the total amount of intracellular Ca^{2+}. In heart muscle, the free Ca^{2+} present during EC coupling does not normally saturate the reaction sites on troponin. Hence, changing the intracellular free Ca^{2+} concentration alters the extent of the Ca-troponin reaction and the disinhibition of actin by tropomyosin. Increasing the free cytoplasmic $[Ca^{2+}]$ increases actin's reaction with myosin, increasing the muscle's contractile force (a positive inotropic effect). Reducing the free cytoplasmic $[Ca^{2+}]$ has a negative inotropic effect, that is, it reduces contractility or inotropic state (IS).

Increasing the inotropic state (IS) increases the contractile performance (rate of tension development and shortening, peak shortening, and rate of relaxation are all increased and the duration of contraction is decreased; see Heart Muscle 1), of heart muscle at each end diastolic fiber length, whereas decreasing the IS decreases the contractile performance. The contractile performance of the heart, of course, is determined by the performance of heart muscle and therefore exhibits equivalent changes. A reduction in IS decreases the contractile activity of the heart, resulting in a decrease in the pumping action.

Extracellular Ca^{2+} concentration could be reduced by a decreased dietary intake of calcium, a decreased vitamin D production, or a defect in the plasma $[Ca^{2+}]$ regulation system.

PROBLEM INTRODUCTION

The contractile performance of heart muscle is determined by its inotropic state, preload and afterload. These factors also determine the stroke output of the heart. The inotropic state is influenced by a variety of factors, including the extracellular Ca^{2+} concentration.

Predict the consequences (increase, decrease, no change) of the acute onset of hypocalcemia on the following.

CARDIAC MUSCLE	PREDICTION
1. velocity of shortening	DECREASE
2. inotropic state (contractility)	DECREASE
3. peak force developed	DECREASE
HEART	PREDICTION
4. stroke volume	DECREASE
5. end systolic volume	INCREASE
6. end diastolic volume	INCREASE

CARDIAC MUSCLE

1. velocity of shortening

   Hypocalcemia (reduced blood Ca^{2+} concentration) reflects a lower than normal extracellular Ca^{2+} concentration. This condition would decrease the Ca^{2+} electrochemical potential of myocardial cells and reduce the passive Ca^{2+} entry during the plateau phase of the cardiac muscle action potential (reduced slow inward current). As a consequence, the muscle cells would have a low IS and a reduced velocity of shortening. See Figure htmus 1-1.

2. inotropic state (contractility)

   Contractility is another term for inotropic state (IS), and it depends on the free intracellular $[Ca^{2+}]$ during excitation contraction coupling. It is reduced by hypocalcemia.

3. peak force developed

   Reduced IS (reduced contractility) decreases the peak force developed at every end diastolic fiber length. See Figure htmus 1-1.

HEART

4. stroke volume

   Hypocalcemia reduces the inotropic state of the heart muscle. As a consequence, at a given end diastolic volume (EDV determines ED fiber length) and a given aortic pressure (afterload), the ventricular stroke output is smaller.

5. end systolic volume

   With a reduced stroke volume, the volume of blood remaining in the ventricle at the end of ejection, the end systolic volume (ESV), is greater.

6. end diastolic volume

   Initially, the venous return is unaffected. Hence, an unchanged filling added to a larger ESV results in an increased ventricular volume at the end of filling, or increased EDV. The increased EDV increases ventricular performance (Starling's law of the heart), partially compensating for the smaller IS. In the final steady state, the stroke volume is reduced but is not as small as it would have been at the onset of the hypocalcemia. See Figure htmus 2-1.

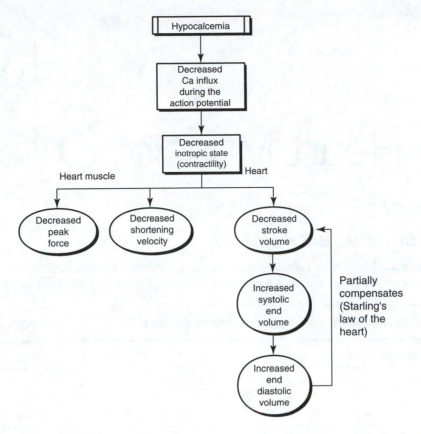

Figure htmus 2-1. **The effects of hypocalcemia on heart muscle and the heart.**

Heart Muscle 3

AFTERLOAD

PHYSIOLOGICAL BACKGROUND

The contractile function of cardiac muscle and therefore the heart depends on three parameters: inotropic state (see Heart Muscle 1 and Heart Muscle 2), preload, and afterload (see table below).

DETERMINANTS	HEART MUSCLE	HEART
inotropic state	yes	yes
preload	resting fiber length	end diastolic volume
afterload	additional weight lifted during shortening	aortic pressure and aortic valve outflow resistance

When the afterload is increased, contractions are slower (force velocity relationship) and less shortening takes place. The contractions are also more forceful and are of shorter duration. When the contractile force of an isolated cardiac muscle strip equals the afterload, shortening begins and the force cannot increase further. When the ventricular pressure reaches the value of its afterload (the aortic pressure), shortening begins. *In vivo*, however, the ventricular pressure and the aortic pressure continue to rise as ejection takes place. The limiting ventricular pressure is the isometric (isovolumetric) pressure for the ventricular volume that is present. This pressure is not normally reached.

CLINICAL BACKGROUND

The cross-sectional area of a normal heart valve is extremely large. Thus a normal valve offers very little resistance to flow, and the ventricular afterload is almost completely due to the arterial pressure. When a valve is stenosed (i.e., the valve orifice is reduced in size), however, it imposes a larger outflow resistance that may then become a significant part of the afterload. Pumping against a large outflow resistance slows the contraction, which enables the ventricle to further increase its pressure (force velocity relationship). The limit remains the isometric pressure for the existing ventricular volume, however.

PROBLEM INTRODUCTION

The contractile performance of heart muscle depends on its inotropic state, preload, and afterload. These factors also determine the stroke output of the heart. In vivo, the aortic pressure, the pressure against which the left ventricle must pump, is the main afterload. If the outflow resistance of the aortic valve should increase, however, it too can impose a significant afterload on the ventricle.

Predict the consequences (increase, decrease, no change) of an increase in afterload on heart muscle and the effect of a sudden stenosis of the aortic valve (reduction in the cross-sectional area of the valve) on the heart. Explain your predictions.

CARDIAC MUSCLE	PREDICTION
1. velocity of shortening	DECREASE
2. peak force developed	INCREASE
3. shortening	DECREASE
HEART	PREDICTION
4. end systolic volume	INCREASE
5. end diastolic volume	INCREASE
6. stroke volume	DECREASE

CARDIAC MUSCLE

1. velocity of shortening

   All muscles exhibit a force velocity relationship. The velocity of shortening is inversely related to the afterload. Hence, an increase in cardiac muscle afterload will reduce its shortening velocity.

2. peak force developed

   At each fiber length there is a maximum force—its isometric force—that a muscle can develop. (See Figure htmus 3-1. At length P, the isometric force is i.) For cardiac muscle, the isometric force depends on inotropic state (contractility) as well as the initial muscle length, preload. The effect of inotropic state is not shown in the figure. But a contracting muscle develops as much force as the load it bears, up to the limit of its isometric force. Hence, in this problem, the peak developed force increases if the afterload increases (see Figure htmus 3-1). The afterload before an increase is a_1, and during contraction the developed tension increases from passive tension to a_1. When the afterload is increased to a_2, the fiber develops more tension, from passive tension to a_2.

3. shortening

   Once the muscle force equals the afterload, further crossbridge cycling leads to shortening. The maximum shortening that can be achieved is defined by the length tension relationship for that muscle. Two important determinants come into play: the preload, which determines the starting length, and the afterload, which determines the ending length. The shortening that occurs is the difference between the initial and the final lengths (see Figure htmus 3-1). At afterload a_1, the fiber shortens from P to the intercept of a_1 on the total length/tension (L/T) curve, length l_1.

Afterload a_1 is the maximum isometric tension that the muscle can develop at fiber length l_1. Similarly, at the higher afterload, a_2, the fiber shortens from p to the length at which a_2 is the fiber's isometric tension, that is, length l_2.

HEART

4. end systolic volume

Ventricular systole begins when the ventricular volume is the end diastolic volume. This preload determines the end diastolic fiber length. At this preload, the ventricle has the capacity to develop an isovolumetric (isometric) pressure of i, as shown in Figure htmus 3-2. As in the case of the afterloaded cardiac muscle, however, the ventricle contracts iso(volu)metrically only until the ventricular pressure just exceeds its afterload, the aortic diastolic pressure, P_d. Then the ventricular fibers begin to shorten and the ventricle begins to eject blood. As ejection proceeds, the ventricular and aortic pressures increase to a peak value, systolic pressure, and then decline as the ejection rate falls (see the heavy line in Figure htmus 3-2). The magnitude of the pressure rise from P_d is determined by a number of factors, including the EDV, the ventricular contractility, the compliance of the aorta, and the outflow resistance of the aortic valve. Stenosis of the aortic valve increases the outflow resistance against which the ventricle must pump. This increased resistance adds to the afterload, which slows the ventricular shortening and increases the pressure generated by ventricular contraction (see the light line in Figure htmus 3-2). When the ventricular pressure reaches a value corresponding to the isometric pressure for the existing fiber length (ESV), no further shortening is possible. ESV_1 is the fiber length at which the ventricle reaches isometric pressure for the normal contraction, and ESV_2 is the length reached with the increased afterload caused by aortic stenosis.

5. end diastolic volume

With aortic stenosis causing an increased end systolic volume (see Figure htmus 3-2) and venous return (filling) continuing unchanged (initially), the ventricular end diastolic volume will also be increased in cardiac cycles that follow the appearance of an aortic valve stenosis.

6. stroke volume

Stroke volume is to the ventricle what shortening is to cardiac muscle. Thus stroke volume is determined by the preload, which determines the end diastolic volume (EDV), a function of the myocardial fiber length at the end of filling, and the afterload, which determines the final ventricular volume, the end systolic volume (ESV). Stroke volume is the difference between EDV and ESV. Aortic stenosis imposes an increased outflow resistance on the ventricle and increases ESV. It therefore reduces the ventricular stroke output (Figure htmus 3-2, EDV − ESV_2 versus EDV − ESV_1). This action is initially followed by normal venous return during the ventricular filling phase, increasing both the EDV and the ventricular contractile force during subsequent beats (Starling's law of the heart). Hence, the stroke volume is not reduced as much as it would have been without the increase in EDV.

Figure htmus 3-3 shows the sequence of events in aortic stenosis.

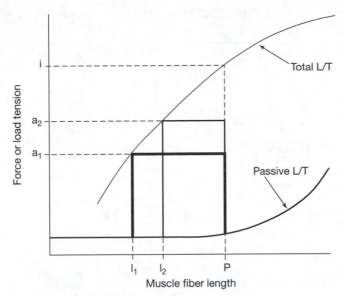

Figure htmus 3-1. The passive and total volume (length, L) pressure (tension, T) relationships for a cardiac muscle preparation. Two afterloaded contractions are shown, starting from the same preload, P. The contraction shown by the heavy line has an afterload of a_1, and the contraction shown by the light line has an afterload of a_2. The fiber lengths at the end of shortening are l_1 and l_2, respectively. The isometric tension at length P is i.

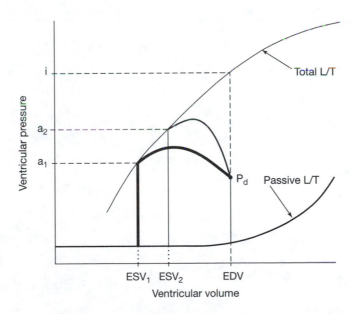

Figure htmus 3-2. Passive and total volume (length, L) pressure (tension, T) curves for the left ventricle. Two systoles are shown, both starting from the same end diastolic volume (EDV). The systole shown by the heavy line is normal, and the one shown by the light line is with aortic stenosis. The normal systole ends at ESV_1, and the systole with aortic valve stenosis ends at ESV_2. The effective afterloads for the two contractions are a_1 and a_2, respectively. The iso(volu)metric ventricular pressure at EDV is i.

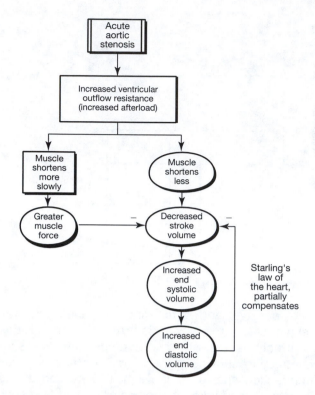

Figure htmus 3-3. The effects of acute aortic stenosis.

Nervous System 1

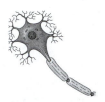

RECEPTOR ORGAN FUNCTION

PHYSIOLOGICAL BACKGROUND

Receptor organs are the initial elements in reflex arcs. They respond to adequate stimuli by changing their membrane potential; the resulting membrane potential change is called the generator or receptor potential (GP). The characteristics of a receptor's response to a stimulus determines the kind of information it can signal to the central nervous system (CNS). Tonic receptors signal the size (strength or amplitude) of a stimulus. Phasic receptors signal a change in the stimulus and how fast that change is occurring. Mixed receptors signal some combination of the two kinds of events. The GP amplitude is usually logarithmically related to the stimulus, allowing receptors to respond to a wide range of stimulus intensities.

The GP in turn determines the action potential (AP) firing rate in the afferent axon. The AP frequency is roughly proportional to the GP amplitude. Generally, the reflex response to a stimulus is determined by the frequency or the pattern of the afferent AP input to the CNS.

PROBLEM INTRODUCTION

An organism's assessment of the external environment and its own internal environment depends on signals that are communicated to its central nervous system. This communication is initiated by a change in the activity of a variety of receptor organs that are responsive to different stimulus modalities.

1. Assume that an adequate stimulus of sufficient intensity to elicit action potentials is applied to a purely <u>phasic</u> receptor organ. The receptor organ responds to the rate of change of stimulus intensity. On the same time axis with the stimulus (S) shown in Figure ns 1-2, plot the generator potential (GP) amplitude and the frequency of action potential firing (F).

The receptor responds only when the stimulus intensity is changing, that is, during the rising and the falling phases of the stimulus. In this instance, during both rising and falling phases the rate of change of stimulus intensity is constant. Hence, the generator potential is constant over each of these intervals. The GP amplitude would be greater for faster stimulus intensity changes, but it would be logarithmically proportional, not linearly proportional to the rate of change. Being purely phasic, the receptor does not respond when the stimulus intensity is constant. APs are produced at a rate proportional to the GP amplitude. APs begin when the GP exceeds a threshold and continue to be generated at a constant frequency (constant GP amplitude) until the GP amplitude falls below the threshold. Once the GP exceeds threshold, the frequency of APs is proportional to the GP amplitude.

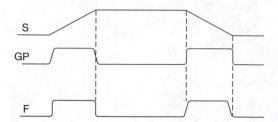

Figure ns 1-2.   S is the amplitude of a stimulus applied to a purely phasic receptor. GP is the generator potential that it produces, and F is the frequency of action potential firing in the afferent neuron.

2. Assume that an adequate stimulus of sufficient intensity to elicit action potentials is applied to a purely <u>tonic</u> receptor organ. The receptor organ response is proportional to the stimulus intensity. On the same time axis with the stimulus (S) shown in Figure 1-4, plot the generator potential (GP) amplitude and the frequency of action potential firing (F).

   This receptor's GP is logarithmically proportional to the stimulus intensity. Once the GP exceeds threshold, APs are produced and continue to be produced for as long as the GP remainsabove threshold. The frequency of APs are directly proportional to the GP amplitude.

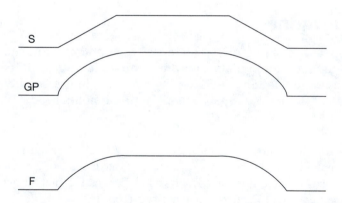

Figure ns 1-4.   S is the amplitude of the stimulus applied to a purely tonic receptor. GP is the generator potential that results, and F is the frequency of action potentials in the afferent neuron. As the stimulus intensity rises linearly, it causes a logarithmically proportional increase in the generator potential. The frequency of action potential firing in the afferent neuron is linearly proportional to the generator potential amplitude.

Nervous System 2

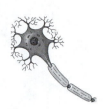

FUNCTION OF A MIXED RECEPTOR

PHYSIOLOGICAL BACKGROUND

The arterial baroceptors are the sensors for the blood pressure-regulating reflex (the baroceptor reflex). Being mixed receptors, they respond to the level of the arterial blood pressure and to the rate of rise of pressure caused by ventricular ejection. The average blood pressure level, the mean arterial blood pressure, is determined by both cardiac function (the cardiac output) and by vascular function (the total peripheral resistance). Acute changes in the rate of rise of pressure are largely determined by changes in the contractile characteristics of the ventricular muscle (changes in cardiac contractility or in ventricular filling). A change in any of these pressure-determining components will affect the baroceptors and hence will affect the baroceptor reflex.

The baroceptor reflex maintains a relatively stable arterial pressure. It does so by modifying the function of the three cardiovascular effectors: the cardiac pacemaker, which determines the heart rate; the myocardium, which determines the pumping characteristics of the ventricles; and the vascular smooth muscle, whose contractile state determines the resistance to blood flow and the distribution of the cardiac output and the blood volume.

CLINICAL BACKGROUND

Coronary artery atherosclerosis (the deposition of fatty plaques in the vessel walls) is a common disorder. It reduces the blood vessel lumen size and limits the blood flow to part(s) of the myocardium either at rest or during increased activity. If the flow limitation is severe enough, it causes chest pain (angina pectoris). The flow limitation can suddenly become complete or almost complete as a result of a clot, hemorrhage, or vasoconstriction (vasospasm) at the site of chronic blockage. If so, the region that is perfused by that vessel becomes ischemic and may die (myocardial infarction). The contribution of that region to cardiac pumping will then be lost.

PROBLEM INTRODUCTION

Moment-to-moment regulation of the arterial blood pressure occurs via the baroceptor reflex. The primary receptors for this reflex, the baroceptors, are stretch receptors in the walls of the major arteries. The receptors in the carotid sinus play a primary role in regulating arterial blood pressure in humans.

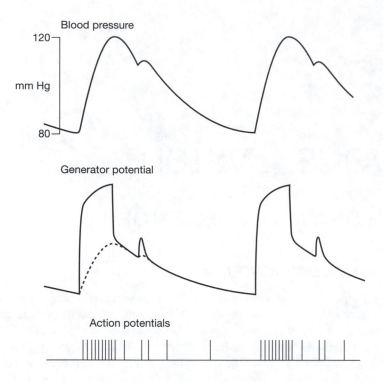

Figure ns 2-3. The generator potential in the carotid sinus baroceptor and the action potentials in the afferent neuron that are caused by a pressure pulse. The dotted line shows the portion of the generator potential that results from its tonic properties. The solid generator potential line shows the results of the combined tonic and phasic properties of the receptor.

1. The arterial blood pressure (BP) is the adequate stimulus for the carotid sinus baroceptor. This organ is a mixed receptor whose response has a tonic component that responds in proportion to the absolute pressure and a phasic component that is sensitive to the rate of pressure rise but does not respond to decreases in pressure. Draw the response of the carotid sinus baroceptors on the same time axis as the arterial pressure recording, shown in Figure ns 2-3. Draw the generator potential (GP) amplitude versus time and the individual action potentials (AP) caused by the GP as single vertical lines.

Figure ns 2-3 shows the magnitude of the generator potential (GP) as a solid line. The contribution of the tonic component to the GP is shown as a dotted line. It parallels the arterial pressure wave. The portion of the GP above the tonic component represents the contribution of the phasic property, which is determined by the rate of change of pressure during the rising phase only. Hence, the phasic component adds to the GP during the initial, rapid pressure rise (the anacrotic limb of the pressure pulse) and during the rising phase of the dicrotic notch, but it does not add to the GP during periods of declining pressure.

The rate of action potential (AP) production parallels the GP. There is an initial rapid burst during the rise to systolic pressure and a second small burst during the dicrotic pressure rise. Between these and during the diastolic period, the GP and hence the AP frequency is determined by the absolute pressure.

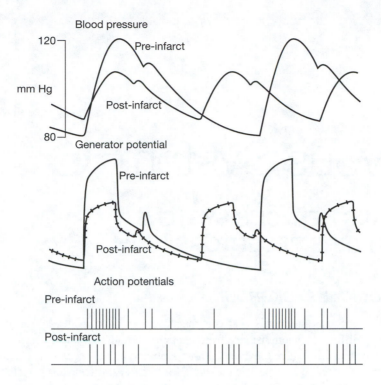

Figure ns 2-4. Two blood pressure tracings, one normal and the other after a cardiac infarct. The middle panel shows the generator potentials that these pressure waves would produce in the carotid sinus baroceptor. The bottom panel shows the afferent neuron action potential firing that would result.

2. Suppose that the individual whose arterial pressure is shown in Question 1 suffered a coronary occlusion that caused a left ventricular myocardial infarction. As a result, his left ventricular ejection rate and his stroke volume were reduced. Reflex sympathetic activation, however, caused vasoconstriction and increased his heart rate. Therefore, his mean arterial pressure was essentially unchanged from normal see Figure ns 2-4, post-infarct BP. Draw the GP response of his carotid sinus baroceptors and the single APs they cause (as individual vertical lines) after the infarct on the same time axis. Both his pre-infarction BP and GP are shown.

The same principles apply to the production of a GP in the baroceptor in this case as in Question 1. Because the <u>rate of pressure rise</u> is greatly reduced by the infarction, there is a smaller contribution by the phasic component. The average pressure over the cycle is not changed, however, so the contribution of the tonic component over an entire cardiac cycle will not be changed. The net effect is a loss of the large depolarization associated with the initial systolic ejection of the ventricle. The small burst associated with the dicrotic notch is also attenuated or lost.

The net effect on AP production is a reduction in afferent impulse traffic to the CNS cardiovascular centers, causing a reflex increase in sympathetic outflow and a reduction in parasympathetic activity. Heart rate will rise, a peripheral vasoconstriction will occur (increase in TPR, total peripheral resistance), and an increase in the inotropic state of the undamaged myocardial fibers will be seen. Without this reflex response, the reduction in functioning myocardial mass would reduce the cardiac output enough to lower the mean arterial pressure.

Nervous System 3

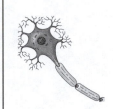

AUTONOMIC NERVOUS SYSTEM: TRANSMITTERS RECEPTORS

PHYSIOLOGICAL BACKGROUND

The autonomic nervous system (ANS) affects the structures it innervates by releasing transmitters. The transmitters diffuse the short distance from the point of release to the innervated cell (effector)—neuron, gland, or muscle—where they react with membrane-bound receptors. The effector will only respond to a transmitter if it has specific receptors for that transmitter. Most effectors have receptors for more than one transmitter. The transmitter-receptor complex triggers a sequence of events that are preprogrammed within the effector cell. Each cell type has a specific response to a transmitter. Thus a particular transmitter activates some effector cells while it inhibits others; or it produces totally different actions in different effectors.

The ANS is generally tonically active. Thus an excitatory neural input to an effector can have a bidirectional influence on that effector. Increased neural activity will further stimulate the effector, whereas inhibition of the neural input will reduce effector activity (less stimulation). The same bidirectional control is exercised by inputs that are tonically active but inhibitory.

PROBLEM INTRODUCTION

The autonomic nervous system (ANS) is widely distributed to and controls the function of the internal organs. The details of that control have largely been elucidated by the use of pharmacological agents that mimic or block the effects of transmitters that are released from the autonomic nerve terminals.

1. An agonist is an agent that reacts with a receptor and mimics the action that normally results from the binding of a transmitter to that receptor. Thus a muscarinic agonist mimics the effects of acetylcholine to a muscarinic receptor. An antagonist blocks the action that normally results from the binding of a transmitter to a particular receptor. Thus a β-adrenergic antagonist blocks the actions of epinephrine (a β-adrenergic agonist) that normally result from its binding to a β-adrenergic receptor.

   Assuming that you had an agonist and an antagonist for every autonomic transmitter receptor, how could you determine which receptor types exist in any autonomically controlled effector?

   An antagonist can only block the effects of transmitters that are being released at the time the drug is administered. Then an antagonist would reduce or eliminate effects that were being produced by the ongoing transmitter secretion and it would reduce or eliminate the effects that the transmitter-receptor complex causes.

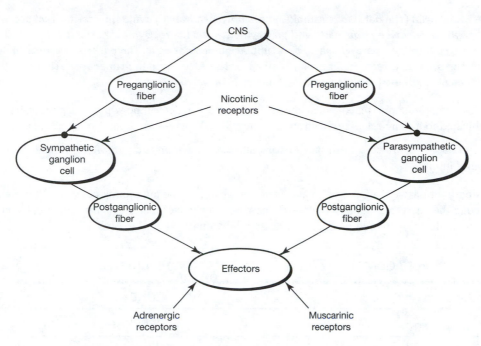

Figure ns 3-1. The location of autonomic transmitter receptors in a schematic diagram of the autonomic nervous system.

If no transmitter were currently being released, however, an antagonist would have no effect. Hence, even if a receptor for that transmitter were present, one would not know.

In contrast, if an effector has a receptor, administering an agonist that binds to the receptor will always produce the same change in activity that the transmitter normally causes. Thus, if one wants to determine whether a particular receptor is present on a cell, one should administer an agonist of the transmitter that normally binds to that receptor.

Alpha-adrenergic agonists generally produce the opposite effects of β-adrenergic agonists when receptors for both are present on an effector cell. Similarly, α-adrenergic agonists generally produce the opposite effects of muscarinic agonists when receptors for both are present on an effector cell. Further, the response to agonists that have the same or opposite actions on an effector may have quite different sensitivities. Nicotinic receptors are located in autonomic ganglia on the postganglionic neurons. Thus nicotinic agonists excite both sympathetic and parasympathetic postganglionic neurons. Because these neurons tend to have opposite peripheral effects, the overall action of a nicotinic agonist may resemble the effects of either sympathetic or parasympathetic stimulation. The response depends on the relative strengths of their actions.

The location of the various receptors is illustrated in Figure ns 3-1.

2. Using the method you defined in Question 1 and your knowledge of the autonomic control of the function of the internal organs, predict the effects (increase, decrease, no change) of the following autonomic agonists on heart rate (HR).

AGONIST	HR CHANGE
α-adrenergic	NONE
β-adrenergic	INCREASE
muscarinic	DECREASE
nicotinic	DECREASE

SA nodal cells (the cardiac pacemaker) have both β-adrenergic and muscarinic receptors and are innervated by both sympathetic and parasympathetic fibers. Sympathetic stimulation increases the heart rate, and parasympathetic stimulation decreases it. The parasympathetic innervation has the greater effect. So, if both divisions of the ANS are simultaneously stimulated, as with a nicotinic agonist (see Figure ns 3-1), heart rate will decrease.

3. Using autonomic pharmacological agents, how could you determine which receptors in an organ are subjected to a tonic release of neurotransmitter?

If the neural input to an effector is tonically active, an antagonist will block the action of the neural input.

4. Using your answer to Question 3 and your knowledge of the autonomic control of heart rate (HR), predict the heart rate changes (increase, decrease, no change) that the pharmacological agents listed in the table below would produce in a resting individual.

ANTAGONIST	HR CHANGE
α-adrenergic	NONE
β-adrenergic	DECREASE
muscarinic	INCREASE
nicotinic	INCREASE

Both the sympathetic and parasympathetic innervations of the SA node are tonically active, but the parasympathetic innervation is more active than the sympathetic. Hence, the increase in HR caused by a muscarinic antagonist is larger than the decrease in HR caused by a β-blocker. A nicotinic blocking agent will eliminate both sympathetic and parasympathetic stimuli. Therefore, the HR will increase (removal of the inhibitory effect of the parasympathetics has a greater effect than removal of the stimulatory effect of the sympathetics), but not as much as by blocking the parasympathetics alone.

5. Predict the heart rate changes that the agents would produce in a vigorously exercising individual.

ANTAGONIST	HR CHANGE
α-adrenergic	NONE
β-adrenergic	LARGE DECREASE
muscarinic	NONE TO SLIGHT INCREASE
nicotinic	INCREASE

The increase in heart rate that occurs during exercise results from a decrease in parasympathetic input to the SA node (which may decrease to zero with vigorous exercise) and an increase in sympathetic input. Hence, blocking the sympathetic effect dramatically reduces HR, whereas blocking the parasympathetic input has little or no effect. Nicotinic block eliminates both, so its effect is equal to the sum of eliminating the sympathetic and parasympathetic effects.

Nervous System 4

HOMEOSTASIS

PHYSIOLOGICAL BACKGROUND

One of the most important and unifying concepts in physiology is that of homeostasis, the maintenance of the constancy of the internal environment of the body. This concept was first clearly enunciated by Claude Bernard, a pioneering French experimentalist of the nineteenth century. His idea was that an organism's ability to maintain homeostasis gave it a degree of independence from the changes that occurred in the external environment. In fact, homeostasis is so central to the function of organisms that the study of physiology largely involves the learning of individual homeostatic mechanisms and how they interact.

PROBLEM INTRODUCTION

The values of many physiological parameters (plasma concentrations, fluid volumes, blood pressure, etc.) are held within very close limits in the body. Maintenance of this constancy is essential to life and is called homeostasis. Homeostasis involves the nervous and endocrine systems, which act as the integrators and the controllers of the processes that bring about homeostasis.

Homeostatic systems all have a specific set of functional components that interact in a limited number of ways to enable each system to carry out its activity. This problem reviews the types of components that must be present and their interactions.

1. Suppose that the plasma concentration of a substance, X, was regulated (i.e., its value was held within some narrow limits). List the components of a control system that would have to be present to carry out this homeostatic activity and the function that those components must perform.

   a. A sensor of $[X]_{plasma}$ or some derivative of it. One cannot regulate a property that is not measured. These sensors are often neural receptor organs, but they may be endocrine gland cells functioning as sensors.

   b. An effector mechanism to add X to the plasma if its concentration falls below the desired level, that is, a SOURCE of X. Without this, one could not correct for a deficit.

   c. A effector mechanism for removing X from the plasma if its concentration rises too high, that is, a SINK for X. Without this, one could not correct for an excess.

   There is another way to achieve the same effect. Instead of having an individually controllable source and sink, X could be lost (or gained) at some constant rate and regulation of $[X]_{plasma}$ could depend only on the rate at which X was added to (or lost from) the plasma.

d. A mechanism for determining what value of $[X]_{plasma}$ is desired, often called the SETPOINT. Without this, the system does not know at what value to hold the regulated parameter.

e. A mechanism for determining the present value of $[X]_{plasma}$ relative to the setpoint, an ADDER (sometimes called a SUMMER or a comparator). This element combines the outputs of the setpoint and the sensor. Without this, the system does not know whether to add X to the plasma or remove it.

f. Finally, one needs a CONTROLLER, a mechanism for adjusting the function of the effectors depending upon whether X needs to be added or removed from the plasma to bring the value of $[X]_{plasma}$ toward the setpoint.

It is not always easy or possible to identify the individual components of a control system. When the components of a system can be identified, however, they correspond with the ones described.

2. Diagram the interaction(s) between the components in your list, showing how these interactions maintain the plasma concentration of X. Add an external disturbance—for example, an increased loss of X from the body in the urine—and show how the system would minimize the effect of this disturbance. In your diagram, use the convention shown in figure ns 4-1.

The components of the control system must be connected together to allow the system to regulate $[X]_{plasma}$ by controlling the activity of the two effectors, the one that adds X to the plasma and the one that removes X from the plasma (Figure ns 4-2).

The arrangement that creates this effect is called negative feedback. For example, if $[X]_{plasma}$ falls too low, the setpoint value is higher than the present value of $[X]_{plasma}$, so their sum (sensor input plus set point) is positive, that is, the controller is stimulated. A larger output from the controller stimulates the source to provide more X and inhibits the loss of X through the sink; that is, more X gets added and less gets taken away, an effect that corrects the error that is present. Similarly, if $[X]_{plasma}$ is higher than desired, negative feedback causes $[X]_{plasma}$ to be lowered.

Note that increased urinary excretion of X is like adding an additional sink for X. It would decrease $[X]_{plasma}$ and cause exactly the same response as the one described above. The normal sink would be inhibited, and the source would be stimulated.

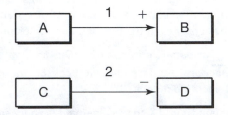

Figure ns 4-1. Panel 1 shows the interaction between two elements in a physiological control system. This representation should be read, "B changes in the same direction as A" or "A acts on B in a direct causal manner." The representation in panel 2 should be read, "D changes in the opposite direction from C" or "C has an inverse causal effect on D. Often, the plus signs are omitted and only the minus signs are shown.

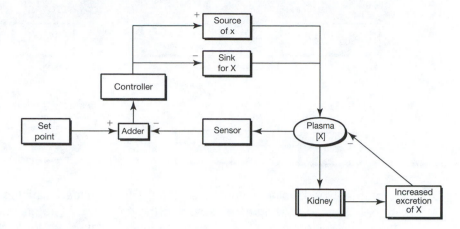

Figure ns 4-2. A generic control system is presented for regulating the plasma concentration of X. It consists of the following functional elements: (1) a sensor of $[X]_{plasma}$, and (2) a setpoint, which are combined by (3) an adder, providing a signal to (4) a controller, which alters the activity of (5) a source and (6) a sink for X, the combined actions of which determine $[X]_{plasma}$. The kidney, as represented here, creates a disturbance in the system by acting as an uncontrolled sink for X.

3. What would be the effect on $[X]_{plasma}$ of blocking the normal interaction (connection) between any two of the components in your system? For example, first eliminate the connection between the controller and between the source and the sink.

Suppose that the controller could not affect the output of X from the source. It would be impossible to correct for the effects of a disturbance. An external disturbance that caused $[X]_{plasma}$ to change would then cause a larger effect than would occur with the control system intact.

In another example, suppose that there was an increased excretion of X in the urine. The $[X]_{plasma}$ would fall. The control system response would inhibit the normal sink and stimulate the source. These changes would reduce or eliminate the effect of increased urinary excretion, and $[X]_{plasma}$ would change very much less than if no control of the source existed.

As another example, suppose that the sensor were disconnected from the adder. Its input, which is normally negative, would be lost and the controller would be stimulated (an unopposed positive input from the setpoint), decreasing the loss of X via the sink and increasing the gain of X from the source. Plasma [X] would rise to very high levels. You can try the effects of any other modifications to the control system and note that some aspect of plasma [X] regulation will be lost.

4. Blood volume is homeostatically regulated by a neuroendocrine mechanism. Invent a neuroendocrine control system that could carry out this regulatory function. Diagram your invention. You need not specify the detailed mechanism by which specific physiological effects are carried out. You need only state that they occur; for example, A acts on organ B to cause effect C. Try out your control system using some conditions you know to affect blood volume.

The control system that you invent must contain all the functional elements that were identified in Question 1: a sensor for the regulated variable, a setpoint, an adder, and a controller, which determines the activity of one or more effectors. There is more than one way to construct such a system. In humans, nature has selected a specific system, part of which is shown in Figure ns 4-3.

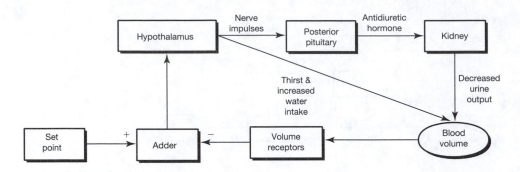

Figure ns 4-3. Part of the blood volume control system. The regulated variable, blood volume, is sensed by volume receptors in the atria. The hypothalamus acts as the controller, changing both the source (water intake) and the sink (renal water excretion).

Let's use diabetes insipidus to test the control system. In this condition, the posterior pituitary cannot secrete an adequate amount of antidiuretic hormone. Urine output increases, lowering the blood volume. The volume receptor output decreases, reducing the negative input to the ADDER. The hypothalamus is stimulated more. The input from the hypothalamus to the posterior pituitary cannot correct the problem because the pituitary cannot secrete more ADH. The hypothalamus, however, does stimulate thirst and increased drinking. Hence, the subject's excess urinary output is balanced by increased fluid intake.

Nervous System 5

SYNAPTIC TRANSMISSION

PHYSIOLOGICAL BACKGROUND

The arrival of an action potential at a postsynaptic terminal initiates a sequence of events that causes the secretion of neurotransmitter. Depolarization of the presynaptic terminal causes voltage-sensitive Ca^{2+} channels to open. Ca^{2+} enters the terminal, causing transmitter-containing vesicles to fuse with the presynaptic membrane. Transmitter is released into the synaptic cleft and diffuses to the postsynaptic membrane, where it binds to specific membrane-bound receptors. The formation of a postsynaptic transmitter-receptor complex triggers a series of events that results in a postsynaptic potential (the end plate potential in muscle). The postsynaptic potential causes a muscle action potential. At most synapses, multiple excitatory inputs must sum to cause a postsynaptic action potential. The motor neuron endings, however, normally release enough transmitter (acetylcholine, ACh) with each presynaptic action potential to cause a postsynaptic action potential.

CLINICAL BACKGROUND

Botulinus toxin (the toxin produced by *Clostridium botulinum*) is the cause of life-threatening food poisoning. Illness begins 12 to 36 hours after ingesting tainted food. The earliest manifestations are gastrointestinal (nausea, vomiting, abdominal pain). Later, neuromuscular symptoms (muscle weakness) appear. Paralysis may result and may prevent respiration. The muscular effects result from the binding of the toxin to presynaptic sites at the neuromuscular junction. This action prevents the exocytotic release of ACh from vesicles and thus prevents the activation and contraction of muscle.

PROBLEM INTRODUCTION

Throughout the nervous system, the predominant mode of neuron-to-neuron and neuron-to-effector (muscle or gland) transmission involves the conversion of an electrical event in the presynaptic neuron (an action potential) to a chemical event (secretion of neurotransmitter) and then back to an electrical event (an action potential) in the postsynaptic cell. Interference with any step in this sequence has significant effects on nerve and or effector function.

Botulinus toxin causes paralysis. It binds to presynaptic sites at the motor end plate (neuromuscular junction) and prevents the exocytotic release of transmitter from vesicles. Predict the consequences (increase, decrease, no change) of blocking the release of acetylcholine at the neuromuscular junction on the following.

PARAMETER	PREDICTION
1. Presynaptic resting potential	NO CHANGE
2. Presynaptic action potential amplitude	NO CHANGE
3. Presynaptic Ca^{2+} influx	NO CHANGE
4. Presynaptic transmitter release	DECREASE
5. Postsynaptic resting potential	NO CHANGE
6. Postsynaptic end plate potential	DECREASE
7. Postsynaptic action potential amplitude	NO AP OR NO CHANGE
8. Postsynaptic action potential frequency	DECREASE OR NO CHANGE
9. Muscle contractile strength	DECREASE OR NO CHANGE

1. Presynaptic resting potential

The size of the resting potential is determined by resting ion permeabilities and the intracellular and extracellular ion concentrations, principally those of K$^+$. A change in the presynaptic transmitter release only affects postsynaptic structures and functions. It has no effect on the resting potential of the presynaptic neuron.

2. Presynaptic action potential amplitude

Action potential amplitude is determined by the E_{Na} and the number of voltage-dependent Na$^+$ channels that open during the regenerative phase of the action potential. Changes in the presynaptic transmitter release affect events that occur subsequently, but do not change presynaptic factors like the presynaptic action potential amplitude.

3. Presynaptic Ca^{2+} influx

Calcium influx into the presynaptic terminals results from the opening of voltage-sensitive Ca^{2+} channels on the arrival of an action potential at the terminals. Neither the presynaptic AP nor the Ca^{2+} channels are altered by the toxin. Hence, presynaptic transmitter release plays no role in presynaptic Ca^{2+} influx.

4. Presynaptic transmitter release

Fusion of transmitter-containing synaptic vesicles with the presynaptic membrane leads to release of transmitter into the synaptic cleft. This process depends on Ca^{2+} entry into the presynaptic terminals. Any condition that interferes with presynaptic action potential generation, Ca^{2+} entry into the presynaptic terminals, vesicle fusion, or transmitter synthesis would reduce the transmitter release. Botulinus toxin does this. It reacts with anchoring proteins in the presynaptic terminals, prevents exocytosis, and reduces the ACh release into the synaptic cleft in the neuromuscular junction.

5. Postsynaptic resting potential

None of the factors that determine resting membrane potential is affected by transmitter release (see the answer to Question 1). Hence, the postsynaptic resting potential, like the presynaptic resting potential, is unaffected by a reduction in transmitter release.

6. Postsynaptic end plate potential

The end plate potential is a localized depolarization caused by the opening of nonspecific ligand-binding cation channels in the postsynaptic membrane. These channels are opened by the binding of acetylcholine to specific membrane receptors. A reduction in transmitter release from the presynaptic terminals will reduce the number of receptor molecules that bind with transmitter, and this will decrease the number of cation channels that are opened. The inward current at the end plate will be reduced; and hence, the depolarization of the end plate will be smaller (a reduced end plate potential).

7. Postsynaptic action potential amplitude

Action potentials are all-or-none phenomena. They either occur or they do not. If the end plate potential is large enough to depolarize the part of the sarcolemma that contains voltage-sensitive Na channels so that the channels open regeneratively (threshold), an action potential of normal size will result. If the end plate potential is too small to depolarize the sarcolemma to threshold, no action potential will occur.

8. Postsynaptic action potential frequency

Normally, each presynaptic action potential releases enough transmitter to cause an end plate potential of sufficient size to trigger one postsynaptic action potential. If a reduced amount of transmitter is released with each presynaptic action potential, the postsynaptic end plate potential will be too small to cause a one-for-one relationship between pre- and postsynaptic action potentials. The frequency of postsynaptic action potentials will be reduced. If the end plate potential is large enough to evoke an action potential, there will be no change in the AP frequency.

9. Muscle contractile force

The strength of skeletal muscle contraction depends on many factors, including the frequency of muscle action potentials. If the generation of muscle action potentials is normal (all other factors remaining unchanged), contractile force will be normal. If not, muscle force will be reduced.

10. Draw a concept map that shows the steps that occur between a presynaptic action potential and the production of a postsynaptic action potential. Show where in this process botulinus toxin exerts its effect.

See Figure NS 5-1.

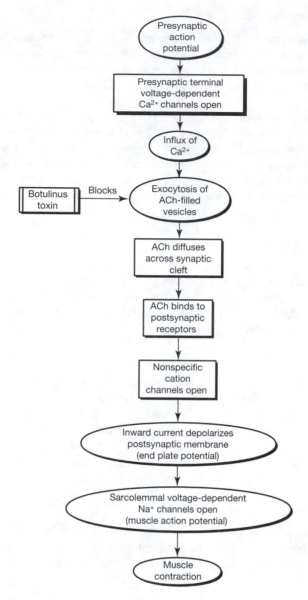

Figure ns 5-1. The steps involved in normal synaptic transmission at the neuromuscular junction are shown. This causal description (cause and effect) of physiological phenomena can be used to predict how the system will function in the face of disturbances or perturbations. The site of action of botulinus toxin is indicated.

Cardiovascular 1

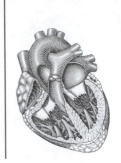

HEMODYNAMICS

PHYSIOLOGICAL BACKGROUND

The resistance to flow in the major arteries and in the major veins is very small. Therefore, the pressure drop along these vessels is small, and the pressure gradient across all organs in a recumbent individual is essentially the same, $P_{\text{aorta}} - P_{\text{vena cavae}}$. Thus differences in the resistances of the various circulatory pathways account for the differences in the blood flow through each path.

When the individual is erect, gravity causes all hydrostatic pressures below the heart to increase and all pressures above the heart to decrease. The postural change does not affect the pressure gradient across the organs below the heart, but it does, however, cause a redistribution of the blood volume. Because of its large compliance, the volume in the peripheral veins increases at the expense of the volume of blood in the central venous compartment. The change in the preload of the heart that this represents ultimately results in activation of the baroreceptor reflex.

PROBLEM INTRODUCTION

The flow of blood in the circulation is described by the laws of hemodynamics. These laws express the relationship between pressure, flow, and resistance. An understanding of the functions of the heart and circulation requires an understanding of these relationships.

Figure cv 1-1 is a model of the cardiovascular system. It consists of a heart and a circulation that perfuses five organs or tissues (1 through 5) as shown. This system *has no reflexes,* and the circulation is assumed to be noncompliant, but in all other ways it behaves like a normal, human cardiovascular (CV) system. The pressure at the aorta (MAP) is maintained constant at 100 mm Hg. Blood flow (Q) through each of the organs or tissues is shown, as are their distances (in centimeters) from the heart.

1. What is the resistance R to flow through organ 3?

   If one knows the pressure gradient ($P_a - P_v$) along a circulatory path and the flow (Q_3) then the fundamental hemodynamic relationship can be used to calculate the resistance to flow (R_3) posed by that pathway:

$$P_a - P_v = Q_3 \times R_3$$

$$R_3 = \frac{P_a - P_v}{Q_3}$$

$$(P_a - P_v) = \text{MAP} - \text{CVP} = 100 - 0 = 100 \text{ mm Hg}$$

$$Q_3 = 2.0 \text{ L/min (see Figure CV 1-1)}$$

$$R_3 = \frac{100}{2} = \underline{50 \text{ mm Hg/L/min/min}}$$

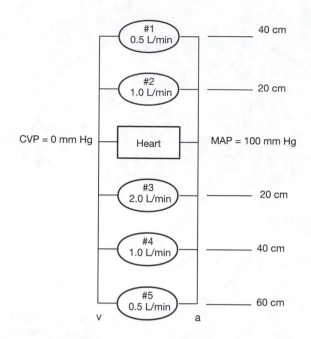

Figure cv 1-1.   A model of a simple cardiovascular system with a heart and five organs or tissues (1 through 5) that are perfused by the heart. This system has no reflexes. The blood flow through each organ is indicated, and the distance of the organ from the heart is shown on the right side.

Similar calculations for the resistances of the other organs would yield

$$R_1 = 200 \text{ mm Hg/L/min}$$

$$R_2 = 100 \text{ mm Hg/L/min}$$

$$R_4 = 100 \text{ mm Hg/L/min}$$

$$R_5 = 200 \text{ mm Hg/L/min}$$

We can thus see that the blood flow through an organ is inversely determined by the organ's flow resistance and that the division of the cardiac output between the organs it perfuses is determined by the relative resistances of the organs.

2. What is the total peripheral resistance (TPR)?

$$\Delta P = CO \times TPR$$

$$TPR = \frac{\Delta P}{CO}$$

$$\Delta P = P_a - P_v = 100 - 0 = 100 \text{ mm Hg}$$

$$CO = Q_1 + Q_2 = Q_3 + Q_4 + Q_5$$

$$CO = 0.5 + 1.0 + 2.0 + 1.0 + 0.5 = 5 \text{ L/min}$$

$$\underline{TPR = \frac{100}{5} = \underline{20 \text{ mm Hg/L/min}}}$$

Alternatively, one can determine the resistance of each path and then calculate the total resistance of the parallel path:

$$R_n = \frac{\Delta P}{Q_n}$$

$$Q_1 = 0.5 \text{ L/min}$$

$$Q_2 = 1.0 \text{ L/min}$$

$$Q_3 = 2.0 \text{ L/min}$$

$$Q_4 = 1.0 \text{ L/min}$$

$$Q_5 = 0.5 \text{ L/min}$$

$$\Delta P = 100 \text{ mm Hg}$$

$$R_1 = 100 \text{ mm Hg}/0.5 \text{ L/min} = 200 \text{ mm Hg/L/min}$$

$$R_2 = 100 \text{ mm Hg/L/min}$$

$$R_3 = 50 \text{ mm Hg/L/min}$$

$$R_4 = 100 \text{ mm Hg/L/min}$$

$$R_5 = 200 \text{ mm Hg/L/min}$$

$$\frac{1}{\text{TPR}} = \frac{1}{R_1} + \frac{1}{R_2} + \frac{1}{R_3} + \frac{1}{R_4} + \frac{1}{R_5}$$

$$= \frac{1}{200} + \frac{1}{100} + \frac{1}{50} + \frac{1}{100} + \frac{1}{200} = \frac{1}{20}$$

$$\underline{\text{TPR} = 20 \text{ mm Hg/L/min}}$$

Note that one effect of arranging the organs in parallel is to decrease the TPR; that is, the TPR is less than any single organ resistance.

The pressure anywhere in the circulation, P, is equal to $P_{\text{pump}} + P_{\text{hydrostatic}}$, where $P_{\text{hydrostatic}} = \rho \times g \times h$. (where ρ is the density of the fluid, g the gravitational constant, and h the height of the column of fluid; $\rho \times g \times h$ for a 1 cm column of blood = 0.75 mmHg).

3. What is the pressure at point a while the individual is lying down? At point v? What is the ΔP across the system $(P_a - P_v)$?

$$P = P_{\text{pump}} + P_{\text{hydrostatic}}$$
$$P_a = P_{\text{pump}} + P_{\text{hydrostatic}}$$

With the individual lying down, there is no vertical column of fluid above points a and v to create a hydrostatic pressure.

$$P_a = 100 \text{ mm Hg} + 0 \text{ mm Hg} \qquad P_v = 0 \text{ mm Hg} + 0 \text{ mm Hg}$$
$$\underline{P_a = 100 \text{ mm Hg}} \qquad \underline{P_v = 0 \text{ mm Hg}}$$
$$\underline{P = P_a - P_v = 100 \text{ mm Hg}}$$

4. With the individual standing quietly, what is the pressure at point a? At point v? What is the ΔP across the system?

With the individual standing, there is a vertical column of fluid above point a and above point v. The pressure at the heart is being held at 100 mm Hg. Points a and v are 60 cm below the heart, and thus there is a 60 cm vertical column present.

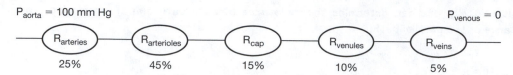

Figure cv 1-2. A model of the flow path of a single organ in the circulation starting with the aorta and proceeding through all the blood vessels in series to the vena cava. The percentage of the total flow resistance of this path represented by each element is indicated.

$$P_a = P_{pump} + P_{hydrostatic} \qquad P_v = P_{pump} + P_{hydrostatic}$$
$$\underline{P_a} = 100 \text{ mm Hg} + 60 \text{ cm} \times 0.75 \text{ mm Hg/cm} = 100 + 45 = \underline{145 \text{ mm Hg}}$$
$$\underline{P_v} = 0 \text{ mm Hg} + 60 \text{ cm} \times 0.75 \text{ mm Hg/cm} = 0 + 45 = \underline{45 \text{ mm Hg}}$$
$$\underline{P} = P_a - P_v = 145 - 45 = \underline{100 \text{ mm Hg}}$$

Thus, although the absolute pressures are increased below the heart in an erect individual, the pressure gradient across any circulatory pathway below the heart is unchanged.

Figure cv 1-2 represents the anatomical elements making up one of the organ circulations in the previous model; the relative resistance to flow posed by each vascular component is indicated.

5. What is the pressure at the distal end of the arterioles?

The pressure drop to any point, X, along the pathway, $P_a - P_x$, depends on R from the origin (aorta) to that point (R_{a-x}), the total R along the path (R_T; i.e. R from "a" to "v"), and the total pressure gradient ($P_a - P_v$).

$$(P_a - P_x)/(P_a - P_v) = R_{a-x}/R_T \quad \text{OR}$$
$$P_x = (1 - R_{a-x}/R_T)(P_a - P_v) + P_v$$
$$P_x = (1 - .70/1.00)(100 - 0) + 0$$
$$P_x = 30 \text{ mm Hg} = \text{pressure at distal end of arterioles}$$

6. What will happen to the pressure at the midpoint of the capillary (P_{mc}) if P_{aorta} is increased by 10 mm Hg? If P_{venous} is increased by 10 mm Hg?

$$P_{mc} = \left(1 - \frac{R_{a-mc}}{R_T}\right)(P_a - P_v) + P_v$$

$$P_{mc} = 1 - \frac{(0.25 + 0.45 + 0.075)}{1.00} = 0.225(P_a - P_v) + P_v$$

If $P_a = 100$ and $P_v = 0$

$$P_{mc} = 22.5$$

However, if $P_a = 110$ and $P_v = 0$ (P_a increased)

$$\underline{P_{mc} = 24.75} \text{ an increase of } (24.75 - 22.5) = \underline{2.25 \text{ mm Hg}}$$

On the other hand, if $P_a = 100$ and $P_v = 10$ (P_v increased)

$$\underline{P_{mc} = 30.25} \text{ an increase of } (30.25 - 22.5) = \underline{7.75 \text{ mm Hg}}$$

Raising P_v increases P_x and P_{mc} more than raising P_a, which means that an increase in venous pressure is more likely to lead to edema (net filtration out of the capillaries) than is an increase in arterial pressure.

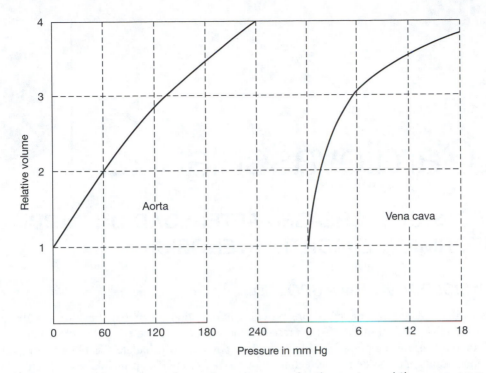

Figure cv 1-3. Pressure-volume (compliance) curves for the aorta and the vena cava.

When the volume contained in an elastic structure undergoes a change (ΔV), the pressure within it changes (ΔP). Conversely, if the pressure within is changed, there will be a corresponding change in volume. The relationship between the change in pressure and the change in volume for an elastic structure is called the <u>compliance</u> (C), which is defined as $\Delta V/\Delta P$.

Figure cv 1-3 illustrates the pressure-volume relationships for the aorta and the vena cava.

7. Calculate the compliance for each type of vessel at normal P_{artery} and normal P_{vein}.

The normal range of pressure in the aorta is 60–120 mm Hg. The normal range of pressure in the vena cava is 0–6 mm Hg. Therefore,

$$C_{aorta} = \frac{\Delta V}{\Delta P} = \frac{0.8}{60} = 0.013 \text{ L/mm Hg}$$

$$C_{vena\ cava} = \frac{\Delta V}{\Delta P} = \frac{2}{6} = 0.33 \text{ L/mm Hg}$$

Note that in their normal pressure ranges, the vena cava (veins) is very much more compliant than is the aorta (or the arteries). Once the veins are distended, however, they have a lower compliance.

Cardiovascular 2

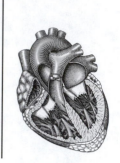

EFFECTS OF INCREASED AFTERLOAD ON CARDIAC PERFORMANCE (AORTIC STENOSIS)

PHYSIOLOGICAL BACKGROUND

The cardiac catheterization procedure provides information about the pressure and the oxygen content (ml O_2/100 ml blood) that is present at the site of the catheter tip. If a measurement of the rate of oxygen consumption is made, it is then possible to calculate the blood flow present at different points in the system. This technique makes use of the Fick principle, a statement of the law of the conservation of matter. It is a simple but extremely powerful principle used by physiologists studying many different systems.

The Fick principle states that if you have a container being filled and emptied, and the contents of the container are constant, then the sum of the inputs to the container must equal the sum of the outputs from the container (see Figure cv 2-1).

The circulation can be viewed as just such a system. The pulmonary circuit has a single inlet, the pulmonary artery, and a single outlet, the pulmonary veins. In addition, there is uptake of oxygen from the alveolar air into the blood perfusing the lungs in the pulmonary circulation. Thus the rate at which oxygen leaves the lungs in the pulmonary veins ($Q \times [O_2]_{pv}$)must equal the rate at which it enters the lungs in the pulmonary artery ($Q \times [O_2]_{pa}$) plus the rate at which it is being taken up from the lungs ($\dot{V}O_2$). Thus we can write

$$Q \times [O_2]_{pv} = Q \times [O_2]_{pa} + \dot{V}O_2$$

or, rearranging,

$$Q \times [O_2]_{pv} - Q \times [O_2]_{pa} = \dot{V}O_2$$

$$Q([O_2]_{pv} - [O_2]_{pa}) = VO_2$$

$$Q = \frac{\dot{V}O_2}{[O_2]_{pa} - [O_2]_{pv}},$$

where Q is the flow (for the lungs this is the cardiac output), $[O_2]_{pv}$ and $[O_2]_{pa}$ are the oxygen contents of blood in the pulmonary vein and pulmonary artery, respectively, and $\dot{V}O_2$ is the rate of oxygen uptake from the lungs.

This relationship can be applied to any part of the circulation. In the systemic circulation, the flow out of the left heart delivers oxygen to the metabolizing tissues at a rate $Q \times [O_2]$aorta. That rate must equal the rate of oxygen use in the tissues (which in steady-state is the same as the rate of oxygen uptake from the lungs) plus the rate at which oxygen is returned to the right heart ($Q \times [O_2]_{mixed\ venous}$).

Thus, as long as we know the oxygen content of the blood entering a structure, the oxygen content of the blood leaving that structure, and the rate of oxygen uptake, we can determine the flow through that structure.

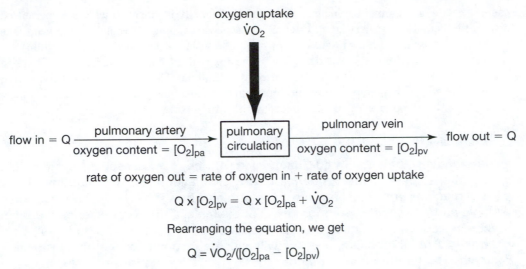

Figure cv 2-1. The Fick principle applied to determining blood flow through the pulmonary circulation.

CLINICAL BACKGROUND

Stenosis of the aortic valve can arise in three ways: (1) congenital malformation of the valve, (2) as a consequence of rheumatic fever, and (3) as a result of calcification in an aging individual.

Individuals with aortic stenosis can be asymptomatic for many years. When the condition becomes more severe, however, it will limit the ability of the individual to exercise. Further increase in severity

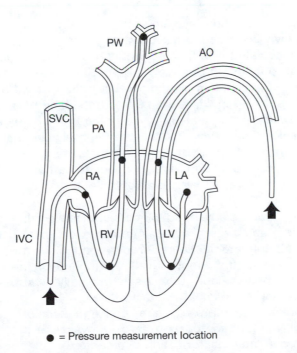

● = Pressure measurement location

Figure cv 2-2. Schematic diagram of the heart and the major vessels. IVC = inferior vena cava; SVC = superior vena cava; RA = right atrium; RV = right ventricle; PA = pulmonary artery; PW = point in pulmonary circulation from which "wedge" pressure is obtained; LA = left atrium; LV = left ventricle; AO = aorta.

causes specific symptoms that include chest pain (essentially identical to what is experienced as a result of ischemia in coronary artery disease) and syncope (fainting). Eventually, the heart is unable to provide the tissues with the perfusion they require to sustain their function (heart failure).

Aortic stenosis is usually treated surgically. The diseased valve is removed, and either a prosthetic valve is put in place or a valve from a pig is used as a substitute. The use of a prosthetic valve requires life-long anticoagulant therapy to prevent clots from forming around the foreign material of the valve. On the other hand, porcine valves tend to deteriorate over a period of years.

PROBLEM INTRODUCTION

Understanding the integrated functioning of the cardiovascular system—the interrelationships between the heart, the circulation, and the cardiovascular control mechanisms—can be aided by determining how this system behaves when a single part of it functions abnormally. For example, if the cross-sectional area of a valve is reduced, the heart will function differently as a pump, there will be a change in arterial pressure, and this will cause a response of the blood pressure regulating system (the baroreceptor reflex). This reflex response will further alter the function of both the heart and the circulation.

Cardiac Catheterization Procedure

Although some information about the function of such an abnormal cardiovascular system can be obtained "from the outside" (systolic and diastolic pressures with a sphygmomanometer, heart sounds with a stethoscope), much of the information needed to characterize the system is available only from within it. Such information has been gained from experimental studies on animal preparations in which lesions to the cardiovascular system have been produced. Now, however, information about cardiovascular abnormalities is also routinely available from medical diagnostic procedures carried out on human patients where nature has performed the "experiment."

Cardiac catheterization is a common procedure used to diagnose cardiac disease. This invasive procedure is used only when serious disease is suspected, but the risks associated with this procedure have in recent years become minimal. A cardiologist obtains information about the state of a patient's heart by introducing a catheter (a thin plastic tube) into the circulation and from there into the heart. The patient is awake but sedated; the electrocardiogram is monitored continuously (yielding the heart rate), as is the rate of oxygen consumption. The catheter is inserted into a vein, usually the femoral vein in the groin, and is slowly advanced into the right atrium, the right ventricle, the pulmonary artery, then finally "wedged" in a small pulmonary artery. A second catheter is threaded through the femoral artery and aorta into the left ventricle. It may be passed still further into the left atrium.

As the catheters are advanced, blood samples are taken and analyzed for their oxygen content, and pressure measurements are made at different locations within the circulation and the heart. A radio-opaque dye may be injected to permit X-ray visualization of the silhouette of the lumina of the blood vessels or the chambers of the heart. Figure cv 2-2 schematically illustrates the locations from which measurements are usually obtained.

In this problem, you will examine the consequences of a single specific cardiac abnormality on the function of the total cardiovascular system using information obtained from cardiac catheterization of a patient.

Mr. A.B., a 56-year-old patient, was admitted to the hospital complaining that he had recently begun to tire on exertion (a round of golf) and that he had occasional heart palpitations. He did not appear to be in acute distress at the time of examination. Physical examination revealed a moderate systolic ejection murmur.

Mr. A.B. was referred for cardiac catheterization, and the following data were obtained:

MEASUREMENT	VALUE
heart rate	76/min
ventilation	5.16 L/min
O_2 consumption	219 ml/min
hemoglobin	13.6 gm/100 ml blood (gm %)
O_2 content: mixed venous blood	13.1 ml O_2/100 ml blood (vol %)
O_2 content: pulmonary artery	13.1 vol %
O_2 content: pulmonary vein	17.4 vol %
O_2 content: aorta	17.4 vol %

1. Calculate Mr. A.B.'s left ventricular output.

During the cardiac catheterization procedure, the patient's rate of oxygen consumption ($\dot{V}_{O_2}$) is continuously measured. As the catheter is advanced through the circulation and the heart, small blood samples are obtained from known locations, and the oxygen content of these samples is measured ($[O_2]$ arterial or venous). With this information, it is possible to apply the Fick principle (see Figure cv 2-1) to determine the blood flow through either the systemic or the pulmonary circulation.

The output of the left ventricle is the systemic blood flow (perfusing the tissues of the body). To determine the systemic blood flow using the Fick principle, one needs values for the oxygen content of blood entering the systemic circulation in the aorta ($[O_2]_{Ao}$) and the oxygen content of the mixed venous blood ($[O_2]_{mv}$) leaving the systemic circulation and returning to the heart.

$$Q_{lv} = \frac{\dot{V}_{O_2}}{[O_2]_{Ao} - [O_2]_{mv}}$$

$$= \frac{219 \text{ ml } O_2/\text{min}}{(17.4 \text{ ml } O_2/100 \text{ ml blood}) - (13.1 \text{ ml } O_2/100 \text{ ml blood})}$$

$$= \frac{219 \text{ ml } O_2/\text{min}}{4.3 \text{ ml } O_2/100 \text{ ml blood}}$$

$$\underline{Q_{lv}} = 5093 \text{ ml blood/min (a normal value)}$$

2. Calculate Mr. A.B.'s pulmonary blood flow.

The inlet to the lungs is the pulmonary artery and the outlet is the pulmonary veins, thus the oxygen content of blood at each of these sites and the O_2 consumption rate will provide information with which to calculate the pulmonary blood flow using the Fick principle. It is important to remember that in the lungs of an individual in steady state, oxygen is taken up by the blood at a rate that exactly equals the rate at which the tissues are consuming it.

$$\underline{Q_{plm}} = \frac{\dot{V}_{O_2}}{[O_2]_{pv} - [O_2]_{pa}}$$

$$= \frac{219 \text{ ml } O_2/\text{min}}{(17.4 \text{ ml } O_2/100 \text{ ml blood} - 13.1 \text{ ml } O_2/100 \text{ ml blood})}$$

$$= \frac{219}{4.3} = \underline{5093 \text{ ml/min (a normal value)}}$$

As expected, because the two sides of the heart are in series and the individual is <u>in steady state, the pulmonary and systemic blood flows are equal</u>.

3. What is Mr. A.B.'s stroke volume?

The cardiac output (CO) always equals stroke volume (SV) times heart rate (HR). Thus we can use this relationship to solve for the value of the stroke volume (heart rate is available from continuous monitoring of the electrocardiogram).

$$SV = \frac{CO}{HR}$$

$$SV = \frac{5093 \text{ ml/min}}{76/\text{min}}$$

<u>SV = 67 ml</u> (a normal value for a person at rest)

Mr. A.B. was diagnosed as having *a stenosed aortic valve*.

4. Given this pathology, predict the changes in pressure (increase/decrease/no change) that would be measured:

PREDICT PRESSURE CHANGES IN:	INCREASE/DECREASE/NO CHANGE
right atrium (mean)	NO CHANGE
right ventricle (systolic)	NO CHANGE
left atrium (mean)	INCREASE
left ventricle (systolic)	INCREASE
aorta (systolic)	DECREASE

The stenosed aortic valve increases the outflow resistance from the left ventricle, which causes left ventricular pressure to increase and increases the pressure drop across the aortic valve, decreasing the aortic pressure. With increased pressure in the left ventricle, pressure in the left atrium will also increase. Unless the condition gets very bad, however, the pulmonary circulation effectively isolates the right side of the heart from the elevated pressures on the left side. The pressures in the right atrium and right ventricle do not change.

5. Diagram the causal relationships that account for the patient's systolic ejection murmur.

The patient's systolic murmur is the result of the increased velocity of flow through the narrowed aortic valve. If the velocity of flow exceeds a critical value, turbulence results, and the energy dissipated is audible as a sound called a murmur. See Figure cv 2-3.

6. Explain the patient's "normality" at rest and his tiring upon exertion.

The stenosed valve creates an increased left ventricular outflow resistance, causing a decrease in aortic pressure. This, of course, activates a baroreceptor reflex that results in increased autonomic stimulation of the heart (increasing heart rate and inotropic state) and increased stimulation of arteriolar smooth muscle (increasing total peripheral resistance). The response of the heart to the increased sympathetic stimulation maintains the cardiac output so that the tissues are adequately perfused when the patient is at rest.

This chronic sympathetic stimulation of the heart, however, causes it to use a part of the reserve cardiac function normally available to support exercise. This action limits the heart's ability to increase its output during exercise, a condition referred to as decreased cardiac reserve. As a result, although cardiac output is adequate when the patient is at rest, when the patient attempts to exercise the output cannot be increased to a level needed to adequately supply the exercising muscles with the oxygen and nutrients they require. The result is fatigue (the patient tires on exertion).

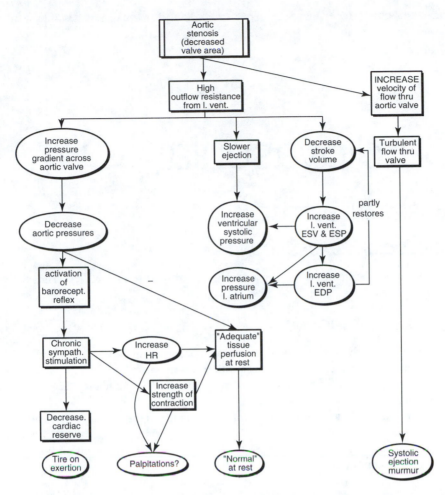

Figure cv 2-3. A concept map of the causal relationships for the physiology present in a patient with aortic stenosis.

Figure cv 2-3. is a causal diagram describing the physiology acting to produce the patient's signs and symptoms. Figure htmus 3 is a diagram of the consequences of aortic stenosis on the myocardium itself.

UNIT 5

Cardiovascular 3

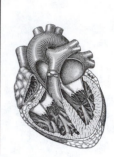

ATRIAL-SEPTAL DEFECT

PHYSIOLOGICAL BACKGROUND

See the answer to Cardiovascular 2 for a description of the use of the Fick principle to determine blood flow.

CLINICAL BACKGROUND

Congenital cardiac defects are not at all uncommon, occurring in 8-10 in 1000 births. They can arise from a number of sources: autosomal chromosome abnormalities, single mutant genes, viral infection during pregnancy or from certain drugs taken during pregnancy. For some reason, these defects are more common in females than in males.

In infants and children, the symptoms may be minor, and the physical signs subtle, making the condition sometimes difficult to diagnose. The occurrence of dyspnea upon exercise later in life may be the first indicator of a congenital defect. Surgical repair of the condition is the only effective remedy for these conditions.

PROBLEM INTRODUCTION

Blood flow through the normal heart is unidirectional, primarily because of the one-way valves in the system. Failure of the valves to function normally or the presence of abnormal flow paths in the heart or its associated vessels alters this normal pattern of flow and results in serious disturbances to normal cardiovascular function. In this problem, we consider an individual with such pathology.

See Cardiovascular 2 for a description of the cardiac catheterization procedure with which the information about this patient was acquired.

Mrs. J.Y., age 35, was admitted to Presbyterian–St. Luke's Hospital to determine the cause of her complaint of vague chest pains in stressful situations. Angina had been ruled out at another hospital, where she also had an echocardiogram because of the presence of a heart murmur. The echocardiogram suggested the presence of right ventricular enlargement. She was sent to this particular hospital for further evaluation.

Right ventricular enlargement was confirmed, and examination revealed a moderate systolic murmur and a split second heart sound.

Mrs. J.Y. underwent cardiac catheterization, and the following data were obtained:

MEASUREMENT	VALUE
heart rate	85/min
ventilation	5.46 L/min
O_2 consumption	219 cc/min
hemoglobin	13.4 gm/dl
O_2 content: mixed venous blood	13.14 ml/dl
O_2 content: pulmonary artery	16.31 ml/dl
O_2 content: pulmonary vein	17.78 ml/dl
O_2 content: aorta	17.78 ml/dl

1. Calculate Mrs. J.Y.'s pulmonary blood flow.

Mrs. J.Y.'s pulmonary blood flow can be calculated from the available data by the use of the Fick equation:

$$Q_{pulm} = \frac{\dot{V}_{O_2}}{[O_2]_{pv} - [O_2]_{pa}}$$

$$= \frac{219 \text{ ml } O_2/\text{min}}{(17.78 - 16.31) \text{ ml } O_2/100 \text{ ml blood}}$$

$$= \frac{219}{1.47} \times 100 \text{ ml blood/min}$$

$$\underline{Q_{pulm}} = 14,898 \text{ ml/min} = \underline{14.9 \text{ L/min}}$$

2. Calculate Mrs. J.Y.'s systemic blood flow.

You can also calculate Mrs. J.Y.'s systemic blood flow using the Fick equation:

$$Q_{sys} = \frac{\dot{V}_{O_2}}{[O_2]_{Ao} - [O_2]_{mv}}$$

$$= \frac{219 \text{ ml } O_2/\text{min}}{(17.78 - 13.14) \text{ ml } O_2/100 \text{ ml blood}}$$

$$\underline{Q_{sys}} = \frac{219}{4.64} \times 100 \text{ ml blood/min} = 4720 \text{ ml/min} = \underline{4.7 \text{ L/min}}$$

3. Compare the two results and explain how they are possible.

Flow out of the right heart through the pulmonary circulation is very much greater than flow out from the left heart through the systemic circulation. For this to be the case, there must be an abnormal flow path present from the left to the right side of the circulation. This is supported by the increase in O_2 content of the blood between the great veins (mixed venous O_2 content) and the pulmonary artery. These data demonstrate that oxygenated blood must be returning directly to the right heart or pulmonary artery. Figure cv 3-1 illustrates a normal heart and three possible examples of such a "shunt."

The following pressures were recorded during cardiac catheterization:

MEASUREMENT	VALUE
right atrium (mean)	6 mm Hg
right ventricle (systolic/diastolic/end diastolic)	30/2/8 mm Hg
pulmonary artery (systolic/diastolic/mean)	30/6/14 mm Hg
left atrium (mean)	6 mm Hg
left ventricle (systolic/diastolic/end diastolic)	106/-3/10 mm Hg
aorta (systolic/diastolic/mean)	102/66/84 mm Hg

4. From the data above determine the location of the abnormal flow path that is present in this patient.

If a large left-to-right shunt (carrying a flow of more than 10 L/min) is present then the resistance to flow through the shunt must be very small and the difference in pressure that is normally present between the two loci will be diminished by the occurrence of that flow.

The pressure difference between aorta and pulmonary artery does not seem particularly diminished, so it is not likely that a patent ductus arteriosus is present.

The pressure difference between left ventricle and right ventricle does not seem to have become smaller, making it unlikely that there is a ventricular septal defect.

The pressures in the atria are normally very small with LA pressure measurably greater than RA pressure. However, in this case, RAP is increased and the two pressures are equal. We can conclude that it is most likely that there is an atrial septal defect in this patient (see Figure cv 3-1). This would usually be verified by injecting a radiopaque dye and observing radiologically the dye flow between the atria.

Mrs. J.Y. has an atrial septal defect, a hole in the septum separating the two atria.

5. Given this pathology, explain the following signs and symptoms:

vague chest pains in stressful situations

systolic murmur

split second heart sound

right ventricular enlargement

"normal" MAP

The presence of a significant left atrium-to-right atrium shunt (atrial-septal defect) has significant consequences for both the left and the right sides of the heart. The right ventricle becomes dilated (enlarged, but not necessarily hypertrophied) by the abnormally large volume of blood that fills it. Pulmonary blood flow is increased by the increased output of the right ventricle (Frank-Starling effect). If the flow through the pulmonary artery is large enough turbulence will occur and a systolic murmur will be present.

In addition, more time is required for the right ventricle to eject its enormous stroke volume than for the left ventricle to eject its reduced stroke volume. The result is a split second heart sound (the right semilunar valve closes later than the left semilunar valve).

Because of the shunt filling of the left ventricle is reduced and hence left ventricular (cardiac) output is relatively low. This would result in decreased mean arterial pressure, but a baroreceptor reflex is generated that results in chronic, elevated sympathetic stimulation of the heart and blood vessels. Heart rate is increased resulting in cardiac output falling only a little. Hence, even with the shunt, mean arterial pressure is more or less normal.

However, chronic sympathetic stimulation results in a reduction of the cardiac reserve. The heart has less ability to increase its output under conditions of exercise or stress, with the result that under these conditions coronary blood flow is inadequate and ischemic pain (vague chest pains in stressful situations) result.

See the accompanying Figure cv 3-2 for an explanation of the origin of the five signs and symptoms.

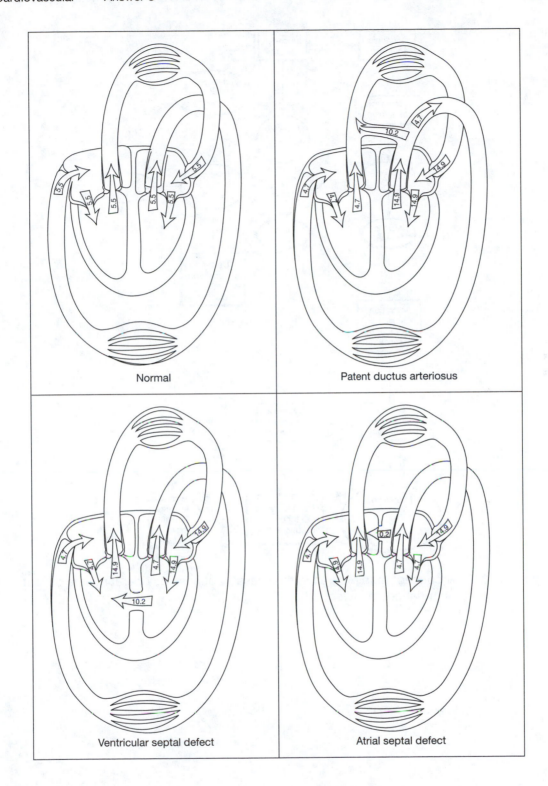

Figure cv 3-1. Schematic views of the normal circulation and representations of the three possible abnormal flowpath (shunts) that could account for the cardiac catheterization data (Richard Levis).

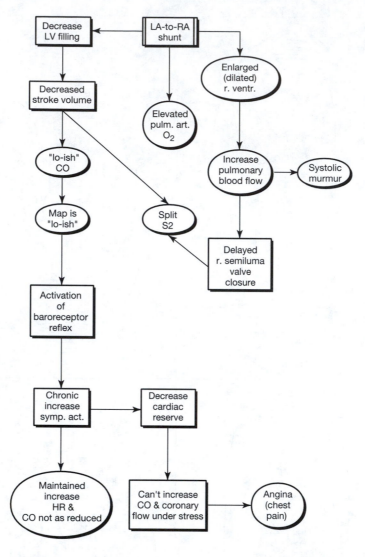

Figure cv 3-2   A concept map illustrating the causal relationships operating in a patient with an atrial-septal defect (LA-to-RA shunt).

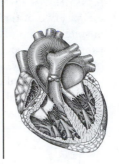

Cardiovascular 4

MITRAL STENOSIS

PHYSIOLOGY BACKGROUND

Cardiac abnormalities not only affect the function of the entire cardiovascular system. Because all organs of the body are perfused by the circulation, any change in cardiovascular function can potentially influence the behavior of every part of the body. Not only can the supply of nutrients and oxygen be altered, but changes in hydrostatic pressure in the capillary beds can affect the filtration of fluid. The consequences of this will, of course, vary, depending on the particular tissue bed involved.

In this problem, we examine the consequences of a cardiac abnormality on the functioning of both the cardiovascular and the respiratory systems.

CLINICAL BACKGROUND

Shortness of breath is a relatively nonspecific complaint in that it can arise from a great many primary sources, some involving the cardiovascular system, others involving the respiratory system. "Three pillow" orthopnea refers specifically to a shortness of breath that occurs when an individual lies down and finds that breathing is more comfortable if the head is propped up with pillows. The occurrence of such orthopnea along with the evidence of peripheral edema (in feet or hands) suggests, but does not prove, the existence of a problem that is producing pulmonary edema.

In this problem, we examine the consequences of a specific cardiac abnormality, damage to the mitral valve, and its effects on the function of the respiratory system. (Although aortic valve damage results in many similar symptoms, aortic stenosis does not lead to respiratory involvement.) Such valve damage is most commonly the result of rheumatic fever, although calcified atherosclerosis can also result in similar abnormal valve function. Females are more often affected than males.

The data to be considered were obtained from cardiac catheterization (see the answer to Cardiovascular 2 for a discussion of this procedure). In this problem, the data are used to localize (identify) the abnormality and then explain its consequences.

PROBLEM INTRODUCTION

Alterations in function of the cardiovascular system due to an abnormality present in that system frequently give rise to abnormal function in other organ systems. The patient described in this problem presents with just such a mix of signs and symptoms.

See Cardiovascular problem 2 for a description of the cardiac catheterization procedure with which the information about this patient was acquired.

Mrs. V.H. is a 45-year-old, overweight woman who has complained of increasing shortness of breath over the past two or three weeks. She has "three pillow" orthopnea and mild pedal edema. On the day of her hospital admission, she had a cough productive of whitish sputum and a sharp retrosternal pain. On examination, she was found to have diffuse wheezing, associated with acute respiratory distress. Her blood pressure was 180/70 mm Hg, her pulse was 90/min, and her respiratory rate was 26/min. She had a moderate systolic murmur and a diastolic rumble. Her chest X ray showed an enlarged left atrium, pulmonary vascular redistribution (increased filling of her pulmonary blood vessels), and interstitial pulmonary edema. Her ECG showed sinus tachycardia and a suggestion of right ventricular hypertrophy.

Cardiac catheterization was performed, and the following results were obtained:

MEASUREMENT	VALUE
heart rate	82/min
ventilation	11.28 L/min
O_2 consumption	273 cc/min
hemoglobin	14 gm %
(a-v) O_2 difference, pulmonary	7.2 vol %
(a-v) O_2 difference, systemic	7.2 vol %
cardiac output	3.8 L/min
cardiac index	1.8 L/min/M^2
body surface area	2.1 M^2

Pressures measured were:

MEASUREMENT	VALUE
right atrium (mean)	4 mm Hg
right ventricle (systolic/diastolic)	45/5 mm Hg
pulmonary artery (systolic/diastolic/mean)	45/24/31 mm Hg
pulmonary wedge (mean)	18 mm Hg
left ventricle (systolic/diastolic)	110/5 mm Hg
aorta (systolic/diastolic/mean)	109/70/83 mm Hg

1. Calculate the pulmonary vascular resistance (PVR), systemic vascular resistance (SVR), the SVR/PVR ratio.

   To calculate the resistance of any portion of the circulation, you need to know the flow through it and the pressure gradient across it producing the flow. Thus we can write:

$$PVR = \frac{P_{PA} - P_{LA}}{CO} \qquad\qquad SVR = \frac{P_{aorta} - P_{RA}}{CO}$$

$$PVR = \frac{31 - 18}{3.8} \qquad\qquad SVR = \frac{91 - 4}{3.8}$$

$$\underline{PVR = 3.42 \text{ mmHg/L/min}} \qquad \underline{SVR = 22.89 \text{ mm Hg/L/min}}$$

$$\frac{SVR}{PVR} = \frac{22.89}{3.42} = \underline{6.69}$$

PVR and SVR are both high, but SVR/PVR is low (pulmonary resistance is increased more than systemic resistance).

2. Which of Mrs. V.H.'s cardiovascular values are normal? Which are abnormal?

The following values are abnormal: ventilation is high, the a-v O_2 differences are high, cardiac output is low, and all of the pressures on the right side of her heart (RA, RV, PA, and PW) are high.

The presence of a systolic murmur and a diastolic rumble are also abnormal.

3. Propose possible abnormalities to account for these data.

One possible cardiac abnormality that could account for this data is left ventricular failure (inability to maintain a normal output). This would elevate LA pressure and, if high enough, increase the pressures in the pulmonary circulation. Pulmonary hypertension would then give rise to pulmonary edema, decreasing oxygen uptake and thereby giving rise to shortness of breath.

The abnormal heart sounds (the murmur and rumble), however, suggest that there is a problem with flow through a valve on the left side of the heart, either the mitral or the aortic. The systolic murmur could arise from either an incompetent (does not close fully) mitral valve or a stenotic aortic valve; in either case, elevated pressure in the left ventricle during systole gives rise to turbulent flow through the affected valve. The diastolic rumble (or murmur) could arise from either an incompetent aortic valve or a stenotic mitral valve.

Figure cv 4-1 is a recording made during Mrs. V.H.'s cardiac catheterization. The upper tracing is the ECG. The lower tracings show the pulmonary wedge (PW) pressure and the diastolic phase of the left ventricular (LV) pressure; the LV pressure goes off the scale during systole. The two pressures were recorded simultaneously at the same recording sensitivities (deflection/mm Hg).

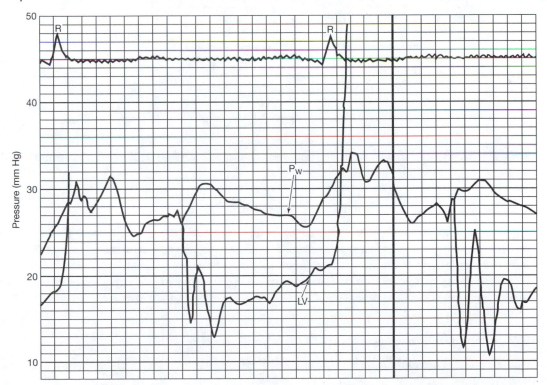

Figure cv 4-1. Tracings obtained during the catheterization of patient Mrs.V.H. The top tracing is the electrocardiogram (ECG). The bottom two tracings are the pulmonary wedge (PW) pressure and the diastolic component of the left ventricular (LV) pressure (at this scale systolic pressure is "off the top" of the page).

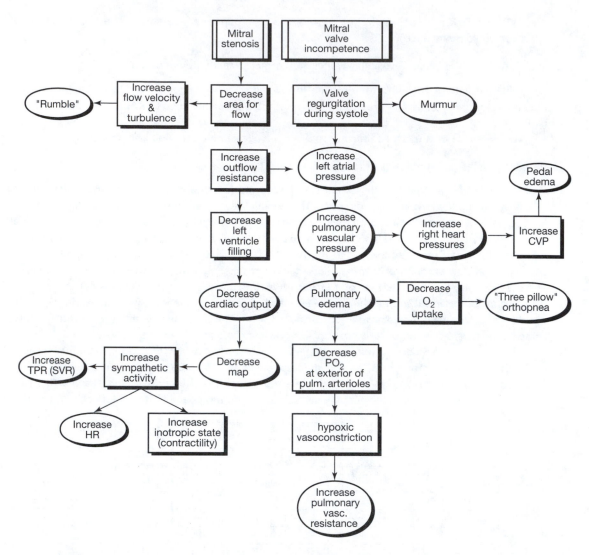

Figure cv 4-2. A concept map illustrating the causal relationships at work in a patient with mitral stenosis.

4. Is any aspect of the above recording (Figure cv 4-1) abnormal?

There is a large pressure difference between left ventricle (LV) and pulmonary wedge (PW) recordings, suggesting increased resistance to flow across the mitral valve. In addition, PW increases still further during systole (when LV pressure goes off the scale), which suggests increased filling of the left atrium due to backflow (or bulging of the mitral valve into the left atrium) across the mitral valve. This situation would give rise to a systolic murmur (but not an ejection murmur).

5. Use the information given here to explain the other abnormalities.

Figure cv 4-2 is a causal diagram outlining the physiology giving rise to the patient's signs and symptoms.

6. What is the cause of Mrs. V.H.'s problem?

Mrs. V. H. has a faulty mitral valve that is both stenosed and incompetent (does not fully close).

7. Construct a causal diagram to represent the physiology of Mrs.V. H.'s cardiovascular system.

Cardiovascular 5

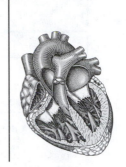

ORTHOSTATIC HYPOTENSION

PHYSIOLOGICAL BACKGROUND

In an erect individual, vertical columns of fluid are present in both the arterial and the venous compartment. Thus the absolute pressure present at any location depends on how far below the heart the vessel is located. Because arterial and venous compartments are equally affected at any distance from the heart, however, the pressure gradient across individual organs or tissues is unchanged.

The increased pressures in the peripheral (in the legs) venous compartment (which is very compliant) results in an increased capacitance there, a reduced volume of blood in the central venous compartment, reduced cardiac filling, and hence reduced cardiac output and mean arterial pressure.

The baroreceptor reflex that results restores mean arterial pressure to near normal by increasing heart rate, cardiac inotropic state, and total peripheral resistance. This response depends on the normal functioning of the autonomic nervous system and the cardiac and vascular effectors.

CLINICAL BACKGROUND

Dysautonomia is a condition in which the autonomic nervous system (ANS) does not function properly. Because of the widespread distribution and effect of autonomic fibers throughout the body, multiple organ systems can be affected. In principle, the defect can occur anywhere along the two neuron pathways that make up the ANS. This condition can be congenital and hereditary or it can be acquired (demyelinating disease, diabetic neuropathy). In either case, often little is known about the origin of the condition.

PROBLEM INTRODUCTION

As the posture of the body changes—for example, in going from lying down to standing up—the effects of gravity on cardiovascular function change significantly. Nevertheless, mean arterial pressure is maintained more or less constantly. In this problem, we consider this phenomenon and the mechanisms that generate it.[1]

1. Consider a normal individual who quickly goes from a supine posture to an erect one. Predict the changes (increase/decrease/no change) that will occur in the following cardiovascular parameters. Compare the values of these parameters while erect with the values of the parameters while lying down.

---

[1]The clinical problem presented here is based on J. E. Eisele, C. E. Cross, D. R. Rausch, C. J. Kurpershoek, and R. F. Zelis (1971), Abnormal respiratory control in acquired dysautonomia, *New England Journal of Medicine*, **285**, 366–368.

PARAMETER	CHANGE WHILE ERECT
stroke volume	DECREASE
heart rate	INCREASE
cardiac output	DECREASE
total peripheral resistance	INCREASE
pressure in root of aorta	DECREASE (SLIGHT)
arterial pressure in foot	INCREASE
venous pressure in foot	INCREASE
capillary pressure in foot	INCREASE
arterial pressure in cerebral blood vessels	DECREASE

When a normal individual, initially recumbent, becomes erect, pressure in all the blood vessels below the heart are subjected to an increased absolute pressure (because of the vertical column of blood above them), and vessels above the heart experience a fall in pressure.

The increase in absolute pressure in all compliant vessels below the heart causes them to increase their volume. The large veins in the legs expand more than the arteries because veins are substantially more compliant than arteries.

The increased volume of blood in the veins of the legs reduces the volume of blood in the central veins. As a consequence, central venous pressure falls and the filling pressures of the ventricles are decreased. The Frank–Starling phenomenon then results in a decreased stroke volume. Therefore, cardiac output must fall, and mean arterial pressure will also fall.

This action elicits a baroreceptor reflex that increases heart rate, cardiac inotropic state, and peripheral resistance. As a result of the sympathetic stimulation of the heart stroke, volume falls less than it would have otherwise, but it is still reduced. Therefore, even though heart rate is up, cardiac output remains reduced from the supine value. TPR is elevated, however. Thus mean arterial pressure is reduced, but only by a small amount. Figure cv 5-1 is a representation of the responses described here.

A 47-year-old teacher goes to her physician complaining of dizziness, fainting, and shortness of breath. She has also experienced episodes of diarrhea.

She appears pale and chronically ill. Her heart rate is 70/min supine (lying down) and 68/min sitting and standing. Her blood pressure is 120/74 mm Hg supine, 98/54 mm Hg sitting, and 84/48 mm Hg standing. Heart sounds and lung sounds are normal. Her abdominal exam and her general neurologic exam are both normal. Routine lab exams are normal.

2. Identify any abnormal data in the description of this patient.

The fall in blood pressure on going from supine to sitting to standing is very much larger than what occurs in a normal individual. The (small) fall in heart rate in going erect is also abnormal; heart rate normally increases under these conditions.

Several studies were performed to evaluate the patient's circulatory control mechanisms.

She was placed on a tilt table, which allows rapid changes in the patient's orientation. Her blood pressure while supine was 102/60 mm Hg. She was then tilted to a 45° heads-up position, and her blood pressure fell to 70/35 mm Hg. There was no change in her heart rate when tilted.

3. Evaluate the results of the tilt test. Is the patient's response normal? What possible mechanism(s) could account for this result?

> Blood pressure should have fallen only slightly, and heart rate should have increased. Here, however, the patient's blood pressure (systolic) fell 30 mm Hg and her heart rate did not change, which is clearly an abnormal response. It could be due to failure of the cardiovascular control system and/or failure of the cardiovascular effectors (sinoarterial, or SA, node, myocardium, vascular smooth muscle) to respond properly.

Failure to maintain a nearly constant blood pressure in changing posture from the supine to the erect is called orthostatic hypotension.

4. List all the functional locations within the cardiovascular system where pathology or dysfunction could lead to failure to activate the baroreceptor reflex upon standing.

> baroreceptors (carotid and aortic)
>
> efferent nerves from receptors
>
> central nervous system cardiovascular centers
>
> autonomic nervous system
>
> > preganglionic fibers
> >
> > ganglia
> >
> > postganglionic fibers
>
> receptors on effector cells
>
> effector cells

Administration of 2 mg of atropine (an antagonist of muscarinic receptors) intravenously did not result in any change in heart rate.

5. Is this response normal? How can you account for this result?

> The sinoatrial node of the heart, the pacemaker, normally receives a large tonic input from parasympathetic (vagal) fibers, which lowers the heart rate. Atropine blocks this input to the SA node cells; therefore, heart rate ought to increase.
>
> Her lack of response to atropine is clearly abnormal. It suggests that the patient's SA node is not receiving, or not responding to, the normal autonomic inputs that determine heart rate, both parasympathetic and sympathetic. If only her parasympathetic input were absent, her resting heart rate would be high, but this is not the case.

Isoproterenol (an α-adrenergic agonist) administration resulted in her heart increasing from 60/min to 118/min.

Stimulated release of catecholamines from sympathetic nerve endings (via administration of tyramine and monoamine oxidase inhibitor) produced a rise in blood pressure.

6. Are these responses normal? What do they tell you?

> These responses are normal. Isoproterenol stimulates the α-adrenergic receptors at the SA node, increasing their rate of firing, and therefore increases heart rate. Similarly, the norepinephrine released from sympathetic nerve endings (and preserved from enzymatic degradation by the presence of a monoamine oxidase inhibitor) will stimulate adrenergic receptors and should lead to an increased blood pressure.
>
> Both responses to pharmacologic challenges suggest that the effectors in the cardiovascular system (the SA node cells, ventricular myocardial cells) are capable of responding normally. Thus it appears that their membrane receptors for autonomic transmitters and the releasable transmitter pool are intact and produce the expected physiological change in function.

7. Where in the system do you believe the patient's problem lies? Why?

The results of administering agonists suggest that the effectors (cardiac muscle and vascular smooth muscle), their membrane receptors for the autonomic transmitters, and the autonomic nerve transmitter pool are intact and normal.

It appears, however, that normal tonic stimulation of the SA node is not present. Blocking the parasympathetic input did not lead to increased heart rate, suggesting that parasympathetic inputs are not present. At the same time, the more or less normal heart rate suggests that the normal sympathetic inputs to the SA node are also not present. (If they were, the lack of parasympathetic inputs would result in a high resting heart rate.)

It would thus appear that this patient's problem lies somewhere proximal to the autonomic nervous system, preventing the normal control signals from reaching the cardiovascular effectors. Exactly where this problem lies cannot be determined more precisely from the available evidence.

This "diagnosis" is strengthened by the fact that this patient has experienced a number of different symptoms, of which all suggest autonomic involvement (her diarrhea, for example). This condition is called dysautonomia, although that label does not identify a specific location for a lesion or dysfunction.

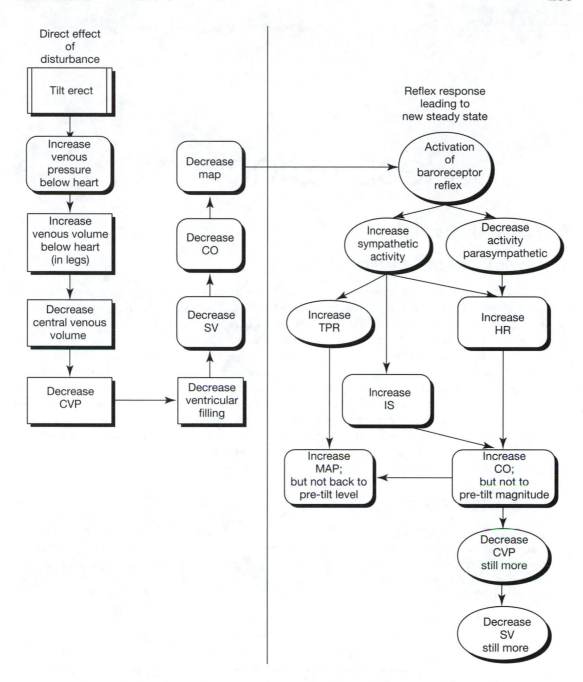

Figure cv 5-1. A concept map describing the cardiovascular responses that occur when an individual is tilted erect from a recumbent position.

Respiration 1

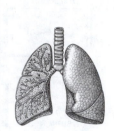

ALVEOLAR VENTILATION

PHYSIOLOGICAL BACKGROUND

With each tidal inspiration, atmospheric air is added to the alveolar air, raising its O_2 content and hence its PO_2. Carbon dioxide, however, is diluted and PCO_2 is reduced by inspiration. With each expiration, part of the alveolar CO_2 is lost to the atmosphere. The gas composition of the alveolar air is determined by this ventilation and the metabolic consumption of O_2 and production of CO_2.

   The effect of ventilation on alveolar gas composition is complicated because only a part of the atmospheric air in each tidal inspiration reaches the alveoli, and alveolar air makes up only a part of the tidal expiration. The last part of each tidal inspiration ends up in nonexchange areas of the lungs—the anatomical dead space—and does not reach the alveoli. The same dead space air, unaltered by gas exchange between the alveoli and the pulmonary capillaries, is the first gas that is expelled from the lungs in tidal expiration. Hence, the effective ventilation, the so-called alveolar ventilation, is the difference between the total amount of gas moved during ventilation and the gas that only ventilates the dead space.

PROBLEM INTRODUCTION

The pattern of ventilation determines the alveolar ventilatory rate, an important determinant of alveolar and arterial gas composition. This problem provides data that may be used to determine the effects of various ventilatory patterns on alveolar gas partial pressures.

The following measurements were made on a normal subject.

PARAMETER	VALUE
P_aCO_2 (mm Hg)	36
P_aO_2 (mm Hg)	100
$\dot{V}_E$ (L/min)	9
$\dot{V}_A$ (L/min)	6
f (breaths/min)	15

DEFINITION OF SYMBOLS	
P_a	arterial partial pressure
$\dot{V}_E$	minute ventilation
$\dot{V}_A$	alveolar ventilation
V_T	tidal volume
V_D	dead space volume
f	breathing frequency

1. Which of the breathing pattern(s) below would not change the subject's P_aO_2? Explain your answer.

PATTERN	V_T (ml)	f (per min)	$\dot{V}_E$ (L/MIN)	$V_T - V_D$ (ML)	$\dot{V}_A$ (L/MIN)
a	400	30	12	200	6
b	500	30	15	300	9
c	800	7.5	6	600	4.5
d	800	10	8	600	6
e	900	10	9	700	7

In a normal person, arterial PO_2 is determined by and is essentially equal to alveolar PO_2. Hence, any pattern of breathing that leaves the alveolar ventilation unchanged would not affect the subject's P_AO_2 or P_aO_2.

The alveolar ventilation for each breathing pattern may be calculated using equation (1):

$$\dot{V}_A = (V_T - V_D) \times f \tag{1}$$

First, we need to determine the value of V_D before we can calculate the values of $\dot{V}_A$ produced by each of the breathing patterns. This value can be done using the subject's original breathing data. The subject's original V_T is determined in the following way:

$$\dot{V}_E = V_T \times f \qquad \text{Rearranging}$$

$$V_T = \frac{\dot{V}_E}{f} \qquad \text{Substituting}$$

$$= \frac{9 \text{ L/min}}{15/\text{min}}$$

$$V_T = \frac{9000 \text{ ml/min}}{15/\text{min}} = \underline{600 \text{ ml.}}$$

Now we can determine the subject's dead space.

$$\dot{V}_A = (V_T - V_D) \times f$$

$$6 \text{ L/min} = (600 \text{ ml} - V_D) \times 15/\text{min}$$

$$V_D = 600 \text{ ml} - \frac{6 \text{ L/min}}{15/\text{min}}$$

$$V_D = 600 \text{ ml} - 400 \text{ ml} = \underline{200 \text{ ml}}$$

We can now complete the table of values for the various breathing combinations using the calculated value for V_D and equation (1) to calculate the subject's alveolar ventilation with each breathing pattern. (See the table p. 00.)

The subject's original V_A was 6 L/min. Breathing patterns "a" and "d" leave alveolar ventilation unaffected. Therefore, they would not cause arterial PO_2 to change. Note that breathing pattern "e" produces the same minute ventilation that the subject originally had. It is not, however, the minute ventilation that determines the value of alveolar (and hence arterial) PO_2, it is the alveolar ventilation.

2. Which breathing pattern(s) would reduce the subject's P_aCO_2 to 24 mm Hg?

Alveolar PCO_2 determines arterial PCO_2, and the two are equal in value in a normal individual. Further, alveolar PCO_2 is inversely proportional to alveolar ventilation rate:

$$P_aCO_2 = P_ACO_2 = K \times \frac{\dot{V}CO_2}{\dot{V}_A},$$

where $\dot{V}CO_2$ = cellular CO_2 production rate and K is a constant to convert to standard temperature. Therefore, to change P_aCO_2 from its original value of 36 mm Hg to a new value of 24 mm Hg would require an alveolar ventilation of 36/24 of the original alveolar ventilation rate: (36/24) × 6 L/min = 1.5 × 6 L/min = 9 L/min. Breathing pattern "b" produces an alveolar ventilation of 9 L/min and would reduce P_aCO_2 to 24 mm Hg.

Respiration 2

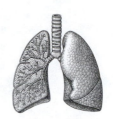

OXYGEN TRANSPORT

PHYSIOLOGICAL BACKGROUND

The presence of hemoglobin (Hb) increases the oxygen carrying capacity of blood 50 times. Because of this, a modest cardiac output (about 5 L/min in humans) is able to deliver enough oxygen to the tissues of warm-blooded species to meet the large demand created by their very high metabolic rate.

The nonlinear relationship between the PO_2 and the oxygenation of Hb contributes importantly to Hb's function as an O_2 carrier. The relatively flat region above a PO_2 of about 60 mm Hg ensures adequate saturation of Hb with O_2 in the lungs, where P_AO_2 is normally about 100 mm Hg. The steep region at PO_2 values below 40 mm Hg ensures that a small reduction in tissue PO_2 will liberate enough O_2 from Hb to meet the tissue needs. Liberating large quantities of O_2 greatly decreases the fall in PO_2 that occurs as O_2 diffuses from the capillary blood into the tissue fluid, which maintains the PO_2 gradient for O_2 delivery to the tissues, protecting them from hypoxic damage.

CLINICAL BACKGROUND

Anemia is a common clinical problem. There are many causes for the condition, such as blood loss, an iron-deficient diet, and inadequate intrinsic factor from the stomach. Each cause has a specific treatment. Whatever the cause, a serious anemia limits the amount of work an individual can do because all anemic individuals have a decreased blood oxygen carrying capacity and a smaller than normal ability to deliver oxygen to the tissues.

PROBLEM INTRODUCTION

Oxygen delivery to the tissues is essential for maintaining their viability and function. One component in this process is oxygen uptake into blood, a process that significantly depends on the oxygen transport protein, hemoglobin. This problem explores the consequences of a hemoglobin deficiency, anemia, on oxygen transport and delivery.

The following questions have to do with identical twins A and B. Twin A is normal, but twin B is anemic. Twin A has a hemoglobin concentration of 15 gm/dl, whereas twin B's is 7.5 gm/dl, one-half that of A. Assume that twin B's anemia has not caused any compensatory changes. An oxyhemoglobin dissociation curve is shown in (Figure resp 2-2). Enter your results in Table 2-1.

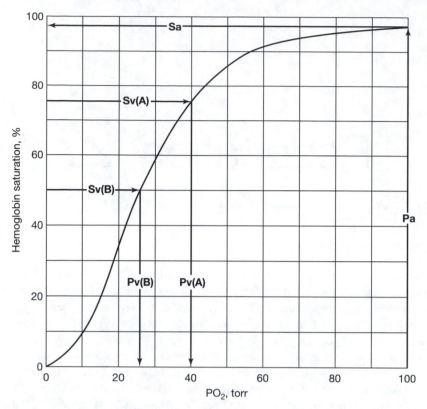

Figure resp 2-2. The oxyhemoglobin dissociation curve shows the relationship between oxygen partial pressure and hemoglobin saturation. Arrows 1 show that a PO_2 of 100 torr (mm Hg) causes a 97% hemoglobin (Hb) saturation. Fifty percent hemoglobin saturation is caused by a PO_2 of 26 mm Hg (arrows 2), and 75% Hb saturation is caused by a PO_2 of 40 (arrows 3).

Both twins have a P_AO_2 (alveolar O_2 partial pressure) of 100 mm Hg.

1. What is twin A's:

a. P_aO_2 (arterial PO_2)?

b. Arterial hemoglobin O_2 capacity (cap HbO_2)?

c. Arterial O_2 content (CaO_2)?

Enter your results in Table 2-1.

a. P_aO_2 (arterial PO_2)

Unless there is a serious impediment to O_2 diffusion, the PO_2 in pulmonary capillary blood equilibrates with the PO_2 in alveolar air. Hence, arterial PO_2 is essentially equal to alveolar PO_2. Twin A's P_aO_2 is 100 mm Hg.

b. Arterial hemoglobin O_2 capacity (cap Hb_{O_2})

This question asks for the maximum amount of O_2 that can be carried by the hemoglobin (Hb) in 100 ml (1 dl) of blood, that is, the amount of O_2 that would be present if the Hb were 100% saturated. It is calculated as the maximum amount of O_2 that can be carried by 1 gm of Hb times the number of grams of Hb in a deciliter of blood.

$$\underline{cap Hb_{O_2}} = cap/gm\ Hb \times number\ of\ grams\ of\ Hb/dl\ blood$$

$$= 1.34\ ml\ O_2/gm\ Hb \times 15\ gm\ Hb/dl\ blood$$

$$= 20.1\ ml\ O_2/dl\ blood$$

Table 2-1

	P_aO_2 (mm Hg)	cap HbO_2 (ml O_2/dl blood)	C_aO_2 (ml O_2/dl blood)
Twin A	100	20.1	20.0
Twin B	100	10.05	10.1

c. Arterial O_2 content (C_aO_2)

Arterial O_2 content per deciliter of blood is the sum of the dissolved O_2 and the O_2 present as HbO_2. The concentration of dissolved O_2 is determined from Henry's law:

dissolved O_2/dl $= PO_2 \times$ solubility of O_2 (ml O_2/dl/mm Hg PO_2)

$= 100$ mm Hg $\times 0.003$ ml O_2/dl/mmHg $= \underline{0.3 \text{ ml } O_2/dl}$

HbO_2/dl $=$ cap $O_2 \times$ % saturated Hb

$= 20.1$ ml O_2/dl $\times 0.98$ (taken from the O_2 dissociation curve at PO_2) $= 100$

(see Figure resp 2-2, arrows P_a and S_a)

$= 19.7$ ml O_2/dl

$C_aO_2 = 0.3$ ml/dl $+ 19.7$ ml/dl $= \underline{20.0 \text{ ml/dl}}$

2. What is twin B's:

      a. P_aO_2 (arterial PO_2)?

      b. Arterial hemoglobin O_2 capacity (cap HbO_2)?

      c. Arterial O_2 content (C_aO_2)?

Enter your results in Table 2-1.

a. P_aO_2 (arterial PO_2)

The blood of the anemic twin also equilibrates with alveolar air in the pulmonary capillaries. Therefore, because both twins have the same P_AO_2, the anemic twin (B) also has a P_aO_2 of 100 mm Hg. A low arterial PO_2 is not characteristic of anemia and is not the source of an anemic individual's problem.

b. Arterial hemoglobin O_2 capacity (capHb_{O_2})

cap $O_2 = O_2$ cap/gm Hb $\times$ number of grams of Hb/dl blood

$= 1.34$ ml O_2/gm Hb $\times 7.5$ gm Hb/dl blood

$= 10.05$ ml O_2/dl blood

    Anemia directly reduces the maximum amount of O_2 that can be carried by the Hb in each deciliter of blood.

c. Arterial O_2 content (C_aO_2)

C_aO_2 ml O_2/dl $=$ amount dissolved/dl $+$ amount present as HbO_2

dissolved O_2/dl $= PO_2 \times 0.003$ ml O_2/dl/mm Hg

$= 100 \times 0.003 = \underline{0.3 \text{ ml/dl}}$

Because the anemic individual has the same arterial PO_2 as a normal person, the concentration of dissolved O_2 is unchanged from normal:

$$\underline{HbO_2/dl} = \frac{cap\ O_2 \times \%\ saturated\ Hb}{100}$$

$$= 10.05 \times 0.98$$

$$= \underline{9.8\ ml\ O_2/dl}$$

The anemic twin has the same P_aO_2 as the normal twin. So, both twins have the same Hb saturation. With half of the normal [Hb], however, the anemic twin has half of the normal HbO_2.

$$\underline{C_aO_2\ ml\ O_2/dl} = dissolved\ O_2 + HbO_2$$

$$= 0.3 + 9.8 = 10.1\ ml\ O_2/dl\ blood$$

The O_2 concentration in an anemic's blood is smaller than normal because of the reduced Hb O_2 carrying capacity.

3. What value of arterial PO_2 would cause twin A's arterial O_2 content to be the same as twin B's?

Twin A's O_2 content is 20 ml O_2/dl, and twin B's is 10.1. Hence , twin A would have to lose 20 − 10.1 = 9.9 ml O_2/dl to have the same O_2 content as twin B. Ignoring the small amount of dissolved O_2, twin A would then have 10.1 ml O_2/dl out of his maximum HbO_2 capacity of 20.1 ml O_2/dl:

$$\frac{10.1}{20.1} \times 100 = \%\ saturation\ of\ Hb = 50\%$$

The PO_2 that yields 50% saturation of Hb (the so-called P_{50}) is 26 mm Hg (see Figure resp 2-2, arrows S_v (B) and P_v (B)).

Therefore, a 75% reduction in PO_2 (a fall from 100 mm Hg to 26) reduces the C_aO_2 to the same extent as a 50% reduction in [Hb]. Losing Hb is more serious than losing an equal percentage of alveolar PO_2. Losing Hb reduces O_2 content in proportion to the Hb loss. Because of the sigmoidal shape of the HbO_2 dissociation curve, however, a decrease in PO_2 produces very little change in O_2 content until the PO_2 falls below about 60 mm Hg.

4. Is it possible for hyperventilation to increase twin B's O_2 content to equal twin A's? Would breathing 100% O_2 do it?

One cannot correct for the loss of Hb by hyperventilating. The best that hyperventilation could do would be to increase the alveolar (and arterial) PO_2 to atmospheric PO_2 (150 mm Hg, correcting for water vapor). Increasing PO_2, however, can only add a small amount of Hb O_2 because twin <u>B's Hb is already 98% saturated</u>. Hence, most of the O_2 would have to be added as dissolved O_2. Raising the PO_2 by 50 mm Hg would only add 50 × 0.003 ml O_2/dl, and this is not enough! In fact, even breathing 100% O_2, which would raise the arterial PO_2 to 760 mm Hg minus the sum of the PCO_2 and the P_{H_2O} [= 760 − (40 + 47) = 673 mm Hg] would only add 573 × 0.003 = 1.7 ml O_2/dl, which is still not enough! It would require a lethal O_2 pressure to dissolve enough O_2 in twin B's blood to make up for the missing Hb.

At rest, both siblings have a cardiac output of 5 L/min and an O_2 consumption of 250 ml/min. (Enter your answers in Table 2-2. p. 293.)

5. What is twin A's:

a. C_vO_2?

b. P_vO_2?

Table 2-2

	C_vO_2 (ml O_2/dl blood)	P_vO_2 (mm Hg)
Twin A	15	40
Twin B	5.05	26

a. C_vO_2

One can use the Fick equation (*see* Cardiovascular 2 for an explanation) to make this calculation:

$$\text{Cardiac output (C.O.)} = \frac{O_2 \text{ consumption rate}}{(C_aO_2 - C_vO_2)}$$

Rearranging yields

$$C_vO_2 = C_aO_2 - \frac{O_2 \text{ consumption}}{C.O.}$$

$$= 20 \text{ ml } O_2/\text{dl blood} - \frac{250 \text{ ml } O_2/\text{min}}{5 \text{ L blood /min}}$$

$$= 20 \text{ ml } O_2/\text{dl blood} - \frac{250 \text{ ml } O_2}{50 \text{ dl blood}}$$

$$C_vO_2 = 20 - 5 \text{ ml } O_2/\text{dl} = \underline{15 \text{ ml } O_2/\text{dl}}.$$

This is an average normal figure.

b. P_vO_2

Ignoring the small amount of dissolved O_2, the venous Hb saturation would be saturation = content/capacity = (15 ml O_2/dl)/(20.1 ml O_2/ dl) = 75%. From the oxyhemoglobin dissociation curve, a 75% saturation is produced by a P_vO_2 of 40 mm Hg [*see* Figure resp 2-2, arrows $S_v(A)$ and $P_v(A)$].

Actually, the PO_2 would be a bit larger if the Bohr effect (the decrease in Hb affinity for O_2 that is caused by the increased PCO_2, increased temperature, and decreased pH of venous blood) were taken into consideration.

6. What is twin B's: (Enter your answers in Table 2-2.)
    a. C_vO_2?
    b. P_vO_2?

One can again use the Fick equation to make this calculation.

$$C.O. = \frac{O_2 \text{ consumption rate}}{C_aO_2 - C_vO_2}.$$

Rearranging,

$$C_vO_2 = C_aO_2 - (O_2 \text{ consumption/C.O.})$$

Twin B's O_2 consumption and C.O. are the same as twin A's so both twins have the same a − v O_2 difference, 5 ml/dl:

10.05 ml/dl − 5 ml/dl = 5.05 ml/dl.

This is very much lower than twin A's venous O_2 content.

b. P_vO_2?

Again we need to first determine the subject's Hb saturation:

$$S = content/capacity = 5.05\ ml/dl/10.05 = 50\%$$

From the Hb dissociation curve, $P_vO_2 = 26$ mm Hg [see Figure resp 2-2, arrows $S_v(B)$ and $P_v(B)$].

7. What do these data tell you about the effects of anemia on oxygen diffusion from blood to the tissues?

Here we can see the problem that faces an anemic individual. If the venous PO_2 is 26 mm Hg, the end capillary blood must have that PO_2 value too. The site of O_2 consumption in cells is in the mitochondria where the PO_2 is very low. Hence, the primary driving force for the diffusion of O_2 is the capillary PO_2 (the gradient = capillary PO_2 − mitochondrial PO_2). A capillary PO_2 of 26 mm Hg may (or may not) be enough to supply resting cells with all the O_2 that they need. It is however, certainly not enough to supply the needs of active cells. Persons with anemia have a reduced capacity to do work. Figure resp 2-3 summarizes the effects of anemia.

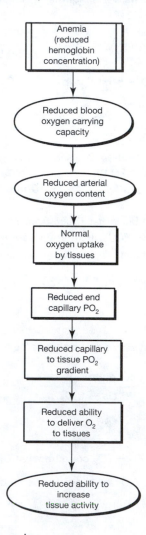

Figure resp 2-3. The effects of anemia.

Respiration 3

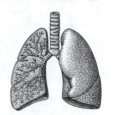

HYPOXIA

PHYSIOLOGICAL BACKGROUND

The barometric (total atmospheric) pressure at sea level is 760 mm Hg, and the pressure decreases with increasing altitude. But each gas in the atmosphere is present as the same fraction (F) of the whole, regardless of altitude. Therefore, each gas, including oxygen, has a smaller partial pressure on Pike's Peak than at sea level ($P_{gas} = P_{total} \times F_{gas}$). Because inspired PO_2 is lower at altitude than at sea level, alveolar and arterial PO_2 will also be lower.

Any change that reduces arterial PO_2 (especially to values below about 60 mm Hg) stimulates the peripheral chemoreceptors. The resulting reflex increases alveolar ventilation, O_2 delivery to the lungs, and the expiration of CO_2. Other disturbances that lower arterial PO_2 also stimulate ventilation via the peripheral chemoreceptors.

CLINICAL BACKGROUND

Many clinical conditions result in hypoxia. In pneumonia, fluid accumulates in the lung interstitium and in the alveoli, interfering with gas diffusion. Tracheal obstruction increases airway resistance. It and CNS (respiratory center) depression therefore reduce alveolar ventilation. A right to left shunt across the lungs allows a part of the venous return to enter the systemic circulation without undergoing gas exchange in the lungs. Beacause each condition is a unique cause, the changes each produces in a variety of physiological parameters differ. The consequences, in terms of tissue oxygenation are the same, however.

PROBLEM INTRODUCTION

Hypoxia is a condition in which the oxygen delivery to the tissues is reduced. There are a number of causes for hypoxia. Although each restricts oxygen delivery, each has a unique set of other characteristics.

1. A healthy college student went to Colorado at the end of the school year for a vacation. While there, he drove to the top of Pike's Peak, a tall mountain. During his first hour on the peak, he walked about but had to sit down to rest three times. The last time was in the weather station. He glanced at the barometer hanging over his head. It read 450 mm Hg.

   Predict the directional changes [increase (I), decrease (D), no change (0)] that would have taken place in each of the listed parameters by the time he entered the weather station compared with the values that he would have had at home (sea level).

	P_AO_2 (mm Hg)	P_aO_2 (mm Hg)	S_a (%)	S_v (%)	V_T (ml)	P_aCO_2 (mm Hg)	pHa
Prediction	D	D	D	D	I	D	I

DEFINITION OF SYMBOLS	
P	partial pressure
A	alveolar
a	arterial
S	Hb saturation
V_T	tidal volume
v	venous

P_AO_2 is determined in part by the inspired PO_2. Hence, lowering the atmospheric PO_2 reduces the alveolar PO_2. Pulmonary capillary blood equilibrates with alveolar air in the lungs so that P_aO_2 and P_AO_2 are normally about equal, that is, P_aO_2 is also reduced. Because hemoglobin saturation is determined by P_aO_2, S_a is also lower. The tissues extract the same amount of O_2 per minute at altitude as they do at sea level. Therefore, the low S_a will be reduced in the tissue capillaries by about as much as would occur at sea level. So S_v and P_vO_2 will also be reduced.

The peripheral chemoreceptors are stimulated more by the low P_aO_2, reflexly increasing both V_T and f. The resulting increase in alveolar ventilation [$V_A = (V_T - V_D) \times f$] increases the rate of expiration of CO_2, reducing the alveolar and the arterial PCO_2. The greater loss of CO_2 causes a respiratory alkalosis, increased pH_a.

2. Predict the effects of the following in an otherwise normal individual at sea level.

	P_AO_2	P_aO_2	S_a	S_v	V_T	P_aCO_2	pH_a
Moderate exercise	O	O	O	D	I	O	O
Pneumonia	I	D	D	D	I	O/D/I	O/I/D
Partial tracheal obstruction	D	D	D	D	D	I	D
CNS Drug depression	D	D	D	D	D	I	D
R to L shunt*	I	D	D	D	I	I/O/D	D/O/I

*R to L lung Shunt = blood flow from right side of heart to left side bypassing the lung.

• Moderate Exercise

During exercise, alveolar ventilation (both V_T and breathing frequency) is increased via a variety of neurological pathways. The chemoreceptor reflexes, however, are not involved, because V_A and metabolism (O_2 consumption and CO_2 production) change proportionately. Because alveolar PO_2 varies directly with O_2 consumption rate and inversely with alveolar ventilatory rate, the alveolar and arterial PO_2 do not change during exercise. Also, because alveolar PCO_2 increases in proportion to CO_2 production rate and decreases in proportion to alveolar ventilation rate, the alveolar and arterial PCO_2 do not change during moderate exercise. With P_aO_2 unchanged, S_a is unaffected. Oxygen consumption, however, increases more than cardiac output so that more O_2 is extracted from each volume of blood perfusing the exercising muscles (increased a - v O_2 content difference). Hence, venous PO_2 and S_v are both reduced.

• <u>Pneumonia</u>

The primary problem in pneumonia is fluid accumulation in the alveolar interstitial space (which increases gas diffusing distance) or in the alveoli (which decreases the gas exchange surface area). Either change decreases O_2 diffusing capacity of the lungs and lowers P_aO_2 and S_a, which causes P_vO_2 and S_v to be lower. The reduced P_aO_2 stimulates the peripheral chemoreceptors and increases V_T and alveolar ventilation and therefore P_AO_2. Because P_ACO_2 varies inversely within alveolar ventilation, it would be decreased. P_aCO_2 decreases with P_ACO_2 but decreases with reduced diffusion distance or exchange surface area. So the P_aCO_2 could be increased, decreased, or unchanged, depending on which of the factors changes most. Finally, arterial pH varies inversely with the PCO_2.

• <u>Partial Tracheal Obstruction</u>

Tracheal obstruction increases airway resistance and interferes with gas flow into and out of the lungs (decreased V_T and decreased $\dot{V}_A$). Hence, it causes P_AO_2 and P_aO_2 to fall and P_ACO_2 and P_aCO_2 to rise. Everything else follows from this: decreased S_a and S_v, and decreased pH_a.

• <u>CNS Drug Depression</u>

The primary problem in drug depression is depression of the brain stem respiratory centers, which causes a reduction in ventilation (V_T and breathing frequency). The consequences are the same as in the case of tracheal obstruction.

• <u>R to L Shunt Across Lungs</u>

The shunted blood has the same composition as mixed venous blood, causing low P_aO_2 and S_a and high P_aCO_2. Reflex hyperventilation (low O_2 stimulation of the peripheral chemoreceptors and possibly high P_aCO_2-related stimulation of the central chemoreceptors) occurs. Hyperventilation, however, cannot make up for the low P_aO_2 because the hemoglobin in the blood that perfuses the lungs is normally saturated to 95–97%. Increasing ventilation cannot add much more O_2 to that blood. The hypoxia-driven hyperventilation increases the CO_2 loss. Therefore, arterial PCO_2 (and pH) depends on how much CO_2 is added by the shunted blood compared with how much extra CO_2 is lost because of hyperventilation. So, P_aCO_2 could be increased, decreased, or unchanged from normal.

Respiration 4

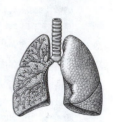

CHRONIC HYPOXEMIA & HYPERCAPNIA

PHYSIOLOGICAL BACKGROUND

Resting ventilation depends importantly on chemoreceptor input to the brain stem respiratory centers. Approximately 80% of the resting ventilatory drive normally comes from the medullary (central) chemoreceptors and about 20% from the peripheral chemoreceptors (carotid and aortic bodies). The properties of these receptors are quite different. The peripheral receptors receive an enormous blood flow per gram of tissue and are rapidly responding. Although the central chemoreceptors are acutely sensitive to changes in arterial PCO_2, they respond slowly because CO_2 must diffuse between the blood and the brain extracellular fluid and react with water to change the local $[H^+]$. The central receptors respond to the change in $[H^+]$.

The receptors differ in other ways. The peripheral receptors are stimulated by a decrease in arterial PO_2, and they do not adapt. The increased central chemoreceptor response to a maintained change in arterial PCO_2 disappears over a period of days. The receptors themselves do not adapt. Rather, there is a mechanism for regulating the cerebrospinal fluid pH. When the $[H^+]$ is restored to normal, the stimulus generated by the change in arterial PCO_2 is gone.

Finally, the response of the peripheral chemoreceptors is not linear. Small decreases in PO_2 provide only a small stimulation. When PO_2 falls below a value of about 60 mm Hg, however, the stimulating effect of low PO_2 increases dramatically. The response of the central chemoreceptors also increases with increasing PCO_2. Above an arterial PCO_2 of about 90 mm Hg, however, CO_2 begins to depress the central nervous system and inhibits ventilation.

CLINICAL BACKGROUND

Chronic obstructive pulmonary disease (COPD) is most commonly the result of emphysema, a condition in which there is destruction of lung tissue. Until recently, COPD had been a disease that predominantly appeared in men from 50 to 60 years in age, but it is now becoming more common in women, related to an increase in female cigarette smoking. The cardinal symptom is breathlessness on exertion, often associated with wheezing. Sufferers are prone to developing bronchitis, especially in the winter.

Emphysema is associated with a decreased lung diffusing capacity and ventilation/perfusion imbalance, leading to chronic hypoxemia. If the condition is severe enough, however, chronic hypercapnia is also present. The hypoxemia and hypercapnia get worse as the disease progresses. When hypoxemia appears, the peripheral chemoreceptor drive increases and becomes more important in determining resting ventilation. Hypercapnia develops slowly, and the central chemoreceptors adapt to it as it gets worse.

PROBLEM INTRODUCTION

The activity of the central and peripheral chemoreceptors determines the resting ventilation. This problem deals with a situation in which these chemoreceptor inputs have been modified by disease.

A 59-year-old carpenter came into the emergency room complaining of shortness of breath. His pulse was 112/min, and his blood pressure 138/88 mm Hg. His respiratory rate was 35/min, and his breathing was extremely labored. Loud breath sounds with some crackling rales were heard over all the lung fields. The patient had an ashen complexion, and his nail beds were cyanotic.

An arterial blood sample was drawn and sent to the lab for blood gas analysis. A chest X ray was ordered, and the patient was given oxygen to breathe.

1. Given the information you have about the patient, predict the deviations from normal (increase, decrease, no change) you would expect him to have in the following parameters *before he was given oxygen*. Explain your predictions.

PARAMETER	PREDICTION
P_aO_2 (mm Hg)	DECREASED
P_aCO_2 (mm Hg)	INCREASED, DECREASED, NO CHANGE
pH_a	INCREASED, DECREASED, NO CHANGE
arterial O_2 content (ml/dl)	DECREASED

P_aO_2 Decreased.

The subject is cyanotic, that is, he has an increased concentration of reduced (deoxy) Hb in his arterial blood, which causes a bluish discoloration of his skin. See Figure resp 4-1. Oxygenation of Hb is determined by PO_2. Cyanosis in this subject results from a low arterial PO_2.

P_aCO_2 All three possibilities need to be considered.

• Increased PCO_2

The patient's description points to the presence of lung disease: shortness of breath, labored breathing, loud breath sounds, and crackling rales. That disease could be associated with a decreased diffusing capacity and/or with a ventilation/perfusion mismatch. Both interfere with the subject's ability to expire CO_2, as they do with his ability to take up O_2, thus causing increased P_aCO_2.

• Decreased PCO_2

The overventilation of normal or less diseased lung areas, induced by hypoxemia acting on the peripheral chemoreceptors, could be great enough to cause an excess loss of CO_2, thus decreasing P_aCO_2.

• No change PCO_2

The combination of increased ventilation of normal lung areas and CO_2 retention in diseased lung areas could result in near normal P_aCO_2.

• pH_a decrease, increase, no change

All three possibilities need to be considered. Changes in pH of respiratory origin vary inversely with the P_aCO_2.

• Arterial O_2 content decreased

Cyanosis is produced by the presence of a higher than normal concentration of reduced Hb in the skin blood vessels. If the concentration of reduced Hb is higher than normal, the concentration of oxyHb must be lower. Hence, total O_2 content, mostly oxyHb, is low. See Figure resp 4-1.

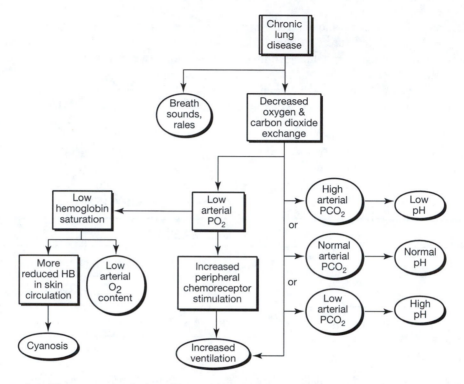

Figure resp 4-1. The causal relationships explaining the signs and symptoms of chronic lung disease.

One-half hour later, the patient was found to be unresponsive. His complexion had changed to a flushed pink with no trace of cyanosis. His respiratory rate was 9/min, and his breathing was quiet. His heart rate was 140/min, and his blood pressure was 85/50 mm Hg. The patient was in a deep coma.

2. Compared with the time the patient came into the ER, predict the changes in the values of the parameters below (increase, decrease, no change) when he became comatose. Explain your predictions.

PARAMETER	PREDICTION
P_aO_2 (mm Hg)	INCREASED
P_aCO_2 (mm Hg)	INCREASED
pH_a	DECREASED
arterial O_2 content (ml/dl)	INCREASED

- P_aO_2 Increased

Increasing inspired PO_2 increases alveolar PO_2 and the alveolar to pulmonary capillary PO_2 gradient. Oxygen diffusion into the pulmonary capillaries in the abnormal lung regions will increase (not much increase is possible in normal lung areas because the Hb is almost completely saturated there). This increase is confirmed by the disappearance of cyanosis, indicating greater saturation of Hb occurring. The skin is now pink.

- P_aCO_2 Increased

The decrease in ventilatory rate and depth of breathing ("ventilation is now quiet") indicates a fall in alveolar ventilation. P_ACO_2 is inversely related to alveolar ventilation rate. Hence, less CO_2 is being expired and more is accumulating. Alveolar and arterial PCO_2 both rise.

- pH_a Decreased

The pH changes inversely with the P_aCO_2

- arterial O_2 content Increased

The cyanosis is now gone and his skin is pink. Hence there is more HbO_2 (and less Hb) than before. This represents an increase the O_2 content.

The results of the blood gas analysis from the blood sample that was drawn *at the time of admission to the ER* were received and are given below.

3. Classify each of the laboratory measurements (high, low, normal).

MEASUREMENT	VALUE	HIGH, LOW, NORMAL
[Hb]	16 gm/dl	HIGH
hematocrit	52%	HIGH
P_aO_2	48 mm Hg	LOW
P_aCO_2	90 mm Hg	HIGH
pH_a	7.35	LOW
[HCO_3^-]	48 mM/L	HIGH

4. Has the patient had his "problem" for:
   a. one hour
   b. one day
   c. one week
   d. one year
   e. more than one year

   Either d or e.

5. What information did you use to arrive at your answer to the previous question?

Hemoglobin concentration and hematocrit are both elevated. The low PO_2 has stimulated renal secretion of erythropoietin, which acted on the bone marrow to increase the production of red blood cells. Although it does not take as long as a year for this to occur, it does take longer than one week. It is likely that the patient has a chronic lung disorder of long standing, which has recently become more serious as a result of an acute lung infection.

Other changes that are present would have occurred in a much shorter time, although it does take several days for the kidneys to produce the increased [HCO_3^-] in response to the respiratory acidosis.

6. When the patient was first admitted, his respiratory rate was 35/min. When he was comatose, however, his respiratory rate was 9/min. What caused the change?

When first admitted, the patient had a significant ventilatory drive from the effect of low P_aO_2 acting on his peripheral chemoreceptors. Placing him in an oxygen tent raised his P_aO_2 enough to remove that stimulus to ventilation. Further, in chronic lung diseases with elevated PCO_2, the central receptors "adapt" to the ventilation-stimulating effect of elevated P_aCO_2. So, the elevated PCO_2 no longer stimulates ventilation excessively. In fact, after O_2 administration, as his ventilation falls, his PCO_2 will increase further (i.e., above 90 mm Hg). At this level, CO_2 begins to narcotize the central nervous system, reducing rather than increasing ventilation.

7. Diagram the sequence of events to show why the patient became comatose.

See Figure resp 4-2.

Figure resp 4-2. The causal relationships explaining the signs and symptoms that result from administering oxygen to a person with severe chronic lung disease.

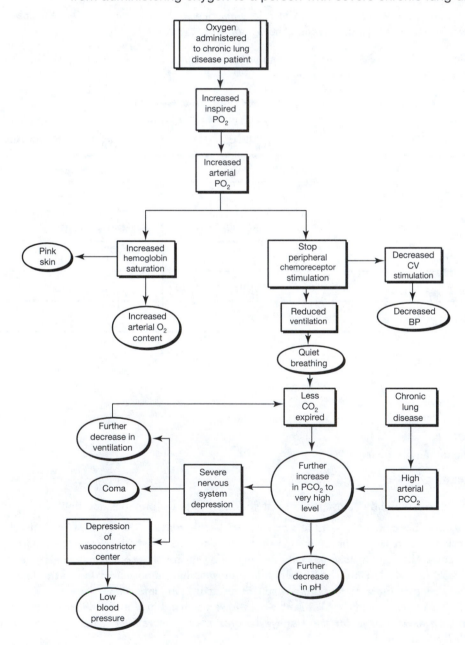

Respiration 5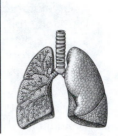

PULMONARY EFFUSION

PHYSIOLOGICAL BACKGROUND

The pumping action of each ventricle establishes a pressure gradient in the vascular circuit it supplies (pulmonary circulation or systemic circulation). An increase in resistance at any point in a circuit (whether it is physiological, such as arteriolar constriction, or pathological, such as a stenosed heart valve) increases the pressure gradient needed to produce a unit of flow through that system. This change is manifested as an increase in pressure upstream of the elevated resistance and a decrease in pressure downstream. In this problem, mitral stenosis increases the resistance to flow out of the left atrium and causes the pressures and volumes in all the upstream structures (in the lungs, right heart, and systemic veins) to increase.

Capillary pressure is the largest component contributing to the filtration of fluid into the interstitial compartment. Under normal circumstances, however, this filtration force is opposed, primarily by the oncotic action of the impermeable plasma proteins. The lungs are normally well protected from edema formation because the pulmonary capillary pressure is quite low. Keeping the lungs dry is particularly important for maintaining gas diffusion. Normally, the alveolar epithelial/capillary endothelial barrier is extremely thin, facilitating gas diffusion. The accumulation of fluid in the interstitium of this barrier can substantially increase the diffusion distance and impede gas transport.

CLINICAL BACKGROUND

Mitral stenosis is the most common cardiac valvular disease. It is frequently caused by rheumatic fever, a streptococcal infection. Among the body regions that are affected by the microorganism are the heart valve leaflets. The growth of the microorganism causes the mitral valve leaflets to stick together, preventing their normal opening. In addition, over a period of many years, the valve may become calcified, causing it to become incompetent (does not close completely) as well.

Stenosis of the mitral valve causes an increase in pulmonary vascular pressures and pulmonary edema. In addition, the abnormal flow through the valve creates flow disturbances that produce heart murmurs. Flow through the mitral valve normally occurs during diastole (during ventricular filling). Hence, stenosis of the mitral valve causes a diastolic murmur. Incompetence of the mitral valve allows backflow through it during ventricular systole, when the valve is normally closed. Hence, an incompetent mitral valve causes a systolic murmur.

The changes in blood gas composition caused by lung edema affects the chemoreceptors. The edema and increased pulmonary blood volume affect lung receptors. Together, they cause the changes in ventilation seen in this patient. The abnormal breathing sounds are caused by reduction in the airways' size so that the high-velocity airflow in them becomes more turbulent.

PROBLEM INTRODUCTION

Physiological systems do not exist in isolation. A perturbation in one system is usually followed by a change in the function of other systems as well. Depending on the disturbance, the effects on other systems may be minimal or quite severe. This problem illustrates an interaction between the cardiovascular and respiratory systems.

Mrs. V.H. is a 45-year-old, overweight woman who was determined, using cardiac catheterization, to have a mitral valve problem (Cardiovascular 4). She has complained of increasing shortness of breath over the past two to three weeks. She has "three pillow" orthopnea and swollen feet. She was admitted to the hospital having a cough productive of whitish sputum and a sharp retrosternal pain. Her respiration was labored, and she had diffuse wheezing.

Her chest X ray showed an enlarged left atrium, increased pulmonary vascular markings (increased filling of her pulmonary vessels), and an indication of interstitial pulmonary edema. Her electrocardiogram showed sinus tachycardia and a suggestion of right ventricular hypertrophy.

On examination her vital signs were:

VITAL SIGN	VALUE
heart rate	90/min
respiratory rate	26/min
blood pressure	115/70
temperature	99.2 °F

1. List the abnormal findings in the patient's description that are indicative of a respiratory problem.

ABNORMAL RESPIRATORY FINDINGS
SHORTNESS OF BREATH
"THREE PILLOW" ORTHOPNEA
PRODUCTIVE COUGH
SHARP RETROSTERNAL PAIN
LABORED RESPIRATION
DIFFUSE WHEEZING
INCREASED PULMONARY VASCULAR MARKINGS
INTERSTITIAL PULMONARY EDEMA
INCREASED RESPIRATION RATE

2. The patient has a mitral valve stenosis (narrowing) and incompetence (doesn't close completely). Draw a causal diagram showing how this condition accounts for the signs and symptoms you listed above.

The causal association of mitral stenosis and regurgitation with the signs and symptoms listed is shown in Figure resp 5-1.

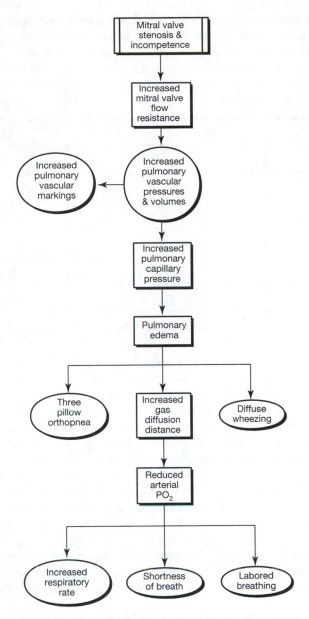

Figure resp 5-1. A concept map showing the causal relations between Mrs. V.H.'s mitral valve disease and her respiratory symptoms.

Mitral stenosis causes an increase in the pulmonary vascular pressures, which in turn cause increased pulmonary capillary filtration and pulmonary edema. The gas diffusion distance then increases, reducing the O_2 diffusion and lowering the arterial PO_2. The low PO_2 stimulates the peripheral chemoreceptors, reflexly increasing respiration rate, and creating the sensation of shortness of breath (labored respiration). The edema in the apex of the lung is reduced by sitting erect or sleeping in a propped-up position ("three pillow" orthopnea). The airways' wall thickness is increased by edema and increased vascular size, narrowing the lumina and causing wheezing. Increased vascular pressure fills apical blood vessels that are normally collapsed, and it stretches blood vessels that are normally open. The source of the increased airways' secretion and the retrosternal pain is uncertain. It may be from a lung infection.

As a consequence of her diseased heart valve, she has pulmonary hypertension.

3. Although it was felt that a major component of her shortness of breath was cardiac in origin, pulmonary function tests were performed to evaluate her for the presence of primary pulmonary disease. Assuming that she does *not* have primary lung disease, predict the changes (increased, decreased, unchanged from normal) you would expect to find in this patient in the parameters listed below.

MEASUREMENT		PREDICTION
Spirometry	Vital capacity (L)	DECREASE
	FEV$_1$ (L)	DECREASE
	FEV$_1$/FVC (%)	UNCHANGED
	MMEF (L/sec)	DECREASE
Lung volumes	FRC (L)	DECREASE
	RV (L)	DECREASE
	TLC (L)	DECREASE
Other	D$_L$CO (cc/min/mm Hg)	DECREASE

Normally, many pulmonary blood vessels are totally or partially collapsed because of the low pulmonary vascular pressures. This patient, however, has elevated pulmonary blood pressure. Therefore, these vessels will be open and distended (pulmonary blood vessels are very compliant). As a result of the increased vascular volume, her lung air space will be reduced. Hence, vital capacity (VC), residual volume (RV), and all the lung capacities that contain RV: functional residual capacity (FRC) and total lung capacity (TLC) decrease.

The increase in vascular volume and interstitial edema cause Mrs. V.H.'s airways to be engorged and their lumina to be narrowed. This increases her airways' resistance, reducing her FEV$_1$ and MMEF. The stiffening of her lungs caused by the vascular engorgement, however, enables her to expel a normal or greater than normal *fraction* of her FVC in 1 second. Hence, the ratio of her FEV$_1$ to her FVC is not reduced. This is similar to what one sees in pulmonary fibrosis where FEV$_1$ and FVC are proportionately reduced.

Mrs. V.H.'s pulmonary diffusing capacity (D$_L$) is determined by the combination of all the factors that contribute to gas diffusion. The thickness of her alveolar epithelial/capillary endothelial diffusion barrier has been increased. Hence, gas diffusion capacity has been reduced.

Her measured pulmonary function data are shown below.

MEASUREMENT		OBSERVED	EXPECTED	OBSERVED/EXPECTED (%)
Spirometry	Vital capacity (L)	3.11	5.09	61
	FEV$_1$ (L)	2.39	3.67	65
	FEV$_1$/FVC (%)	77	72	107
	MMEF (L/sec)	2.11	3.51	60
Lung volumes	FRC (L)	2.06	2.42	85
	RV (L)	0.99	1.24	80
	TLC (L)	4.10	5.94	69
Other	D$_L$CO cc/m/mm Hg	21.98	29.3	75

4. What change (increase, decrease, no change) would you expect to be present in each of the following?

PARAMETER	PREDICTION
arterial PO$_2$	DECREASE
arterial Hb saturation	DECREASE
arterial PCO$_2$	INCREASE, DECREASE, NO CHANGE
arterial pH	DECREASE, INCREASE, NO CHANGE
lung compliance	DECREASE

5. Explain your predictions.

The increased gas diffusion distance prevents the normal PO$_2$ equilibration between alveolar air and pulmonary capillary blood. Thus arterial PO$_2$ is reduced. Hence, hemoglobin saturation is also reduced.

The reduced arterial PO$_2$ increases ventilatory stimulation (increased respiratory rate), which facilitates CO$_2$ loss. The consequence to arterial PCO$_2$ depends on the balance between the effects of the increased ventilation and the effects of the decreased lung diffusing capacity on CO$_2$ loss—increase, decrease, or no change. The pH varies inversely with the change in PCO$_2$.

Lung compliance is reduced because of the stiffening effect of the increased intravascular pressures and the pulmonary edema.

The result of an analysis of Mrs. V.H.'s arterial blood and lung compliance are given below.

PARAMETER	VALUE
ARTERIAL PO$_2$ (MM HG)	74.9
ARTERIAL HB SATURATION (%)	92
ARTERIAL PCO$_2$ (MM HG)	37.9
ARTERIAL pH	7.36
LUNG COMPLIANCE	DECREASED

The measured data confirm the predictions made above.

Renal 1

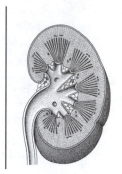

RENAL CLEARANCE AND THE RESPONSE TO OSMOTIC CHALLENGES

PHYSIOLOGICAL BACKGROUND

The kidneys' clearance of a substance, S, is defined as the smallest volume of plasma from which substance S could be totally removed ("cleared") per minute to provide the amount of S that is excreted in the urine each minute. Clearance is given in the units of volume/minute. Thus we can write that the rate at which S leaves the plasma (is cleared from it, or $P_S \times C_S$), must equal the rate at which it appears in the urine (is excreted, or $U_S \times V$):

$$PS \times CS = US \times V \quad \text{or} \quad CS = \frac{U_S}{P_S} \times V.$$

Any substance that is only filtered (i.e., is neither reabsorbed nor secreted) will have a clearance value that equals the glomerular filtration rate (GFR). (Inulin is one such substance, and, in fact, its clearance is used to measure the GFR.) Substances that are filtered and secreted will have clearances greater than the GFR, whereas substances that are reabsorbed will have clearance less than the GFR. Thus comparisons of a substance's clearance with the GFR can give insight into how the kidney handles that substance.

The kidneys' role in maintaining homeostasis is a central one, and disturbances to certain aspects of homeostasis activate mechanisms that change in kidney function, usually in a way that acts to restore homeostasis.

PROBLEM INTRODUCTION

Measurement of the clearance of substances from the blood by the kidneys provides one means of evaluating kidney function. Furthermore, the clearance of certain substances can be used to determine such important parameters as the glomerular filtration rate and the renal plasma flow. The renal clearances of other substances can provide insight into the role of the kidney in maintaining homeostasis.[1]

The data presented on page 309 were collected from a healthy student on three successive days. The data in the column "ad Lib" were collected while the student was allowed to drink water as she chose. The data labelled "Thirsting" were collected after 12 hours of no fluid intake. The "+1 liter Water" data were collected 90 minutes after she drank 1 liter of water. The numbers in the table on the following page that are underlined have been calculated from the data given.

---

[1]This problem is based on Problem 8-1, page 141–143 of H. Valtin. (1967). *Renal Function: Mechanisms Preserving Fluid and Solute Balance in Health.* Boston: Little, Brown, and Company with permission of the publishers.

1. Predict the changes relative to the ad lib state (increase,decrease,no change) that will be present in the data labeled with question marks (??) after Thirsting and after drinking 1 liter of water.

The table below contains the correct predictions.

PARAMETER	AD LIB	THIRSTING	+1 L WATER
V (ml/min)	1.2	0.75	15.0
U_{In} (mg/ml)	15.8	25.2	1.23
P_{In} (mg/ml)	0.151	0.155	0.154
GFR (ml/min)	126	NO CHANGE	NO CHANGE
% filtered H_2O reabsorbed	99.0	INCREASE	DECREASE
P_{Na} (mM)	136	144	134
L_{Na} (mmoles/min)	17.1	17.6	16.1
U_{Na} (mmoles/L)	128	192	10.2
% filtered Na reabsorbed	99.1	NO CHANGE	NO CHANGE
U_{osm} (mOsm/L)	663	1000	100
P_{osm} (mOsm/L)	290	300	287
C_{H_2O} (ml/min)	−1.54	DECREASE	INCREASE
U_{urea} (mg/100 ml)	480	720	48
P_{urea} (mg/100 ml)	12	15	10
C_{urea} (ml/min)	48	DECREASE	INCREASE
% filtered urea reabsorbed	61	INCREASE	DECREASE

2. Now calculate the unknown values (shown as question marks) using the data provided in the table.

The requested calculated values have been inserted in the table below.

PARAMETER	AD LIB	THIRSTING	+1 L WATER
V (ml/min)	1.2	0.75	15.0
U_{In} (mg/ml)	15.8	25.2	1.23
P_{In} (mg/ml)	0.151	0.155	0.154
GFR (ml/min)	126	122	120
% filtered H_2O reabsorbed	99.0	99.4	87.5
P_{Na} (mM)	136	144	134
L_{Na} (mmoles/min)	17.1	17.6	16.1
U_{Na} (mmoles/L)	128	192	10.2
% filtered Na reabsorbed	99.1	99.2	99.1

PARAMETER	AD LIB	THIRSTING	+1 L WATER
U_{osm} (mOsm/L)	663	1000	100
P_{osm} (mOsm/L)	290	300	287
C_{H_2O} (ml/min)	−1.54	−1.75	+9.77
U_{urea} (mg/100 ml)	480	720	48
P_{urea} (mg/100 ml)	12	15	10
C_{urea} (ml/min)	48	36	72
% filtered urea reabsorbed	61	70.5	40

Methods of calculating the requested values

GFR (glomerular filtration rate) = insulin clearance = $\dfrac{U_{In}}{P_{In}} \times V$ (ml/min)

% filtered water reabsorbed = $\dfrac{(GFR - V)}{GFR} \times 100\%$

L_S (filtered load = quantity of S filtered/min) = $GFR \times P_S$ (quantity/min)

Quantity of S excreted /min = $U_S \times V$

% filtered S reabsorbed = $\dfrac{(L_S - \text{quantity excreted/min})}{L_S} \times 100\%$

C_{H_2O} (free water clearance is the rate of clearance of solute-free water)

= rate of water excretion − solute clearance

= $V - \dfrac{U_{osm}}{P_{osm}} \times V$ (ml/min)

$C_{urea} = \dfrac{U_{urea}}{P_{urea}} \times V$ (ml/min)

3. Compare your predictions with the actual values you have calculated. Explain the changes that are present in the thirsting state and after drinking 1 liter of WATER.

Both stimuli (thirsting and drinking 1 liter of water) alter the osmolarity of the body fluids and thus generate an appropriate response from the osmoreceptors. This process alters ADH secretion (from the posterior pituitary) and distal tubule and collecting duct reabsorption of water.

The GFR does not change with thirsting or drinking because the GFR is not controlled to maintain osmolarity constant; the calculated changes seen here are too small to be significant. Further, although a fall in the GFR during thirsting would not help the individual compensate for a change in osmolarity since the filtrate is iso-osmotic to plasma; and a decrease in the GFR following water intake would certainly not help since she must dispose of an increased volume.

The percentage filtered water reabsorbed does, of course, change because this represents the compensation produced by the reflex driven by the hypothalamic osmoreceptors (the rate of secretion of ADH increases when osmolarity increases and is decreased when body fluids are diluted). When osmolarity increases (thirsting), compensation consists of increasing as much as possible the reabsorption of filtered water. The increase from 99.0% to 99.4% represents a 40% increase in reabsorption of the normally unreabsorbed (excreted) water. On the other hand, drinking water (+1 liter) increases the % of filtered water that is excreted by 1200%.

Sodium reabsorption is another parameter that is not altered in these situations, although changes in the sodium load (reflecting a change in the body's sodium level) do elicit a complex reflex (renin release is affected, resulting in increased or decreased angiotensin II, altered aldosterone release from the adrenal cortex, and changes in reabsorption of Na in the distal nephron) to retain or excrete so as to maintain a normal sodium level.

The change in free water clearance follows from the change in the percentage of filtered water reabsorbed. When osmolarity increased, the release of ADH is increased. This release increases the permeability of the collecting duct and results in more water being reabsorbed and a more concentrated urine is being produced because more free water is reabsorbed. When osmolarity is decreased, ADH is decreased, and a more dilute urine is produced because less water is reabsorbed.

The change in urea clearance reflects that urea reabsorption is passive. The reabsorption of water out of the tubule causes the concentration of urea in the lumen to increase. More urea then passively diffuses down the increased concentration gradient from the tubular lumen into the peritubular capillaries. The excretion of urea is thus inversely related to the reabsorption of water.

Renal 2

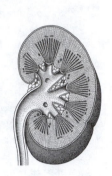

LEAKY GLOMERULUS

PHYSIOLOGICAL BACKGROUND

Kidney function begins with the filtration of water and solutes smaller than 60,000 daltons at the glomerulus. The net filtration that occurs is determined by a balance of hydrostatic and osmotic forces and by the characteristics of the permeability barrier. A change in any of these forces or in the permeability of the barrier will have important consequences for kidney function. Changes in the way the kidneys' handle water and solutes (particularly electrolytes such as sodium) wil then affect the quantity and distribution of fluid between the vascular compartment and the tissues. This change will, in turn, give rise to endocrine responses and changes in the function of the cardiovascular system.

CLINICAL BACKGROUND

One concern about "strep throat" (streptococcal infection) is the possible development of glomerulonephritis, a condition in which glomerular and tubular damage occurs as a result of the immune response to the infection. Glomerulonephritis typically is seen in children and young adults after a latent period of 1 to 4 weeks. Fortunately, most patients (>95%) recover from this condition spontaneously and require only supportive, symptomatic treatment (if any) along with antibiotics to treat the initiating streptococcal infection. During the acute phase of this response, however, several characteristic signs and symptoms, explored in this problem, are seen.

PROBLEM INTRODUCTION

The filtration barrier at the glomerulus is normally permeable to molecules up to approximately 60,000 daltons in size. Thus the plasma proteins do not ordinarily enter the glomerular filtrate. In the presence of pathology that makes the glomerulus "leaky," plasma proteins can be filtered, with serious consequences for renal function, and also for other, functions of the body.

Consider an individual who develops a pathology that results in *only* a leaky glomerulus (permeable to plasma proteins).

1. Predict the changes (increase/decrease/no change) that will be present in the following parameters in the steady state as a consequence of this pathology.

PARAMETER	PREDICTED CHANGE
plasma protein concentration	DECREASE
plasma colloid osmotic (oncotic) pressure	DECREASE
total body sodium	INCREASE
plasma volume	INCREASE
interstitial fluid volume	INCREASE

With a "leaky" glomerulus, plasma proteins will be filtered into Bowman's capsule. These proteins will then be excreted and lost from the body. The decreased concentration of plasma proteins lowers the plasma oncotic pressure. Therefore, there will be increased filtration of water out of the vascular compartment into the interstitial compartment, which has two consequences: (1) the plasma volume will be decreased, and (2) the volume of fluid in the interstitial compartment will be increased. The fluid entering the interstitial contains Na; thus the accumulation of water in the tissues (edema formation) will result in an accumulation of Na and increased total body Na.

As a result of fluid shifts between the vascular and the interstitial compartments, additional changes in function affect cardiovascular and renal function.

2. Predict the changes (increase, decrease, no change) that will occur in the following parameters.

PARAMETER	PREDICTED CHANGE
mean arterial pressure	DECREASE
glomerular filtration rate	DECREASE
renal blood flow	DECREASE
plasma renin concentration	INCREASE
plasma aldosterone concentration	INCREASE
Na excretion	DECREASE
ADH secretion	INCREASE

The decrease in plasma volume results in a decrease in central venous (ventricular filling) pressure, a fall in stroke volume and cardiac output, and hence a fall in mean arterial pressure (MAP). The decreased plasma volume also results in a decreased stimulation of the atrial volume receptors and hence an increased release of antidiuretic hormone (ADH) from the posterior pituitary. This change gives rise to increased retention of water in the kidney and an increase in plasma volume.

The fall in MAP elicits a baroreceptor reflex and increased activity in the sympathetic system. Increased activity in the renal sympathetic nerves has several effects. The renal arterioles constrict the decreasing renal blood flow (RBF) and the glomerular filtration rate (GFR). This action to increased release of renin from the granular cells of the juxtaglomerular apparatus: the fall in pressure in the renal arterioles is sensed by the intrarenal baroreceptor, and the fall in the GFR decreases the delivery of NaCl to the macula densa. At the same time, the increased sympathetic stimulation of the granular cells directly causes release of renin. With renin levels increased, there is increased angiotensin and aldosterone release from the adrenal cortex.

The decrease in Na filtration (GFR is down) coupled with the increase in angiotensin and aldosterone results in increased Na reabsorption (decreased Na excretion).

Figure renal 2-1 is a concept map describing the consequences of "leaky" glomeruli.

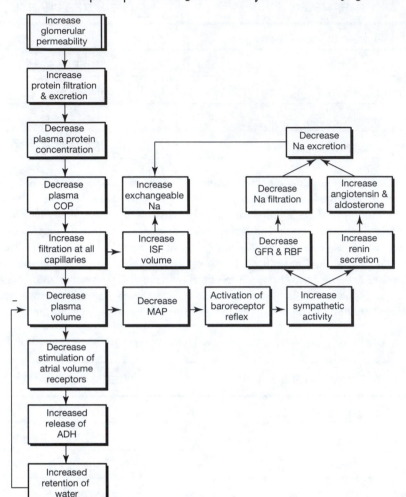

Figure renal 2-1.   Damage to the renal glomeruli, making them "leaky," results in edema, Na retention, and a fall in blood pressure.

An infection with group A streptococcus ("strep throat") can be followed by kidney damage characterized by the development of a leaky glomerulus (increased permeability to plasma proteins) *and* damage to some renal tubules and destruction of others.

3. Predict the initial qualitative changes (increase, decrease, no change) that will occur in the following parameters as a result of this pathology.

PARAMETER	PREDICTED CHANGE
glomerular filtration rate	DECREASE
plasma protein concentration	DECREASE
plasma colloid osmotic (oncotic) pressure	DECREASE
total body sodium	INCREASE
plasma volume	INCREASE
interstitial fluid volume	INCREASE

With the kidney damage described here, the primary insult is the reduced GFR that results from damage to the tubules (filtration cannot occur into Bowman's capsule if flow cannot occur down the tubule). As a consequence, fluid and Na retention result in the appearance of hypertension (from expanded plasma volume). In those nephrons with intact tubules, there will be filtration of plasma proteins (the glomeruli are leaky), a reduction in plasma oncotic pressure, and hence an expansion of the interstitial volume.

As a result of the pathology present and the compensatory mechanisms that are activated, there are additional changes to body function.

4. Predict the changes (increase, decrease, no change) that will occur in the following parameters.

PARAMETER	PREDICTED CHANGE
mean arterial pressure	INCREASE
plasma renin concentration	DECREASE
plasma aldosterone concentration	DECREASE
Na excretion	DECREASE
ADH secretion	DECREASE

MAP is increased because plasma volume is expanded (ventricular filling pressure, or preload, is increased). The reduced GFR results in decreased Na excretion. Na retention and the decrease of renal sympathetic activity result in a decrease in renin release and hence a decrease of aldosterone secretion from the adrenal cortex. The increased plasma volume increases stimulation of the atrial volume receptors, and ADH secretion is reduced.

Figure renal 2-2 on page 316 is a concept map describing the consequences of tubular damage and "leaky" glomeruli.

5. In the first patient, the consequences of a leaky glomerulus leads to a sustained *decrease* in blood pressure. In the second case, in which leaky glomerulus and tubular damage are present, the result is a sustained *increase* in blood pressure. Explain these two quite different outcomes of similar (but, of course, not identical) pathology.

In the first instance, the initiating pathology is a "leaky glomerulus and a consequent fall in plasma volume, which then gives rise to a baroreceptor reflex and widespread increased activation of sympathetic nerves that attempts to compensate for the fall in MAP.

In a situation in which both tubular damage and "leaky" glomeruli are present, the primary disturbances are a fall in GFR and a fall in Na excretion. Na retention then occurs, which expands the plasma volume and leads to hypertension. A leaky glomerulus results in movement of some of the retained fluid into the tissues where edema wil be present.

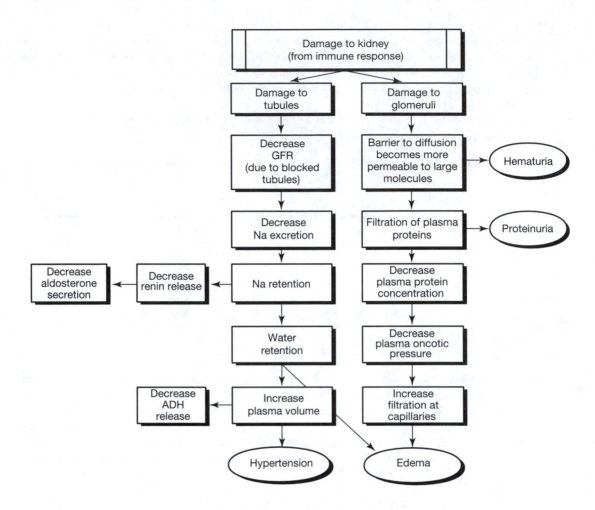

Figure renal 2-2. Kidney damage that extends to both the tubules and the glomeruli leads to fluid and Na retention and to hypertension.

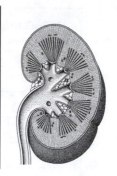

UNIT 7

Renal 3

Hyperaldosteronism

PHYSIOLOGICAL BACKGROUND

Aldosterone controls a kidney's handling of sodium and potassium, promoting the reabsorption of Na and the excretion of K. As K is depleted, the cells of the distal nephron increase their secretion of H^+ (hence increasing the reabsorption of filtered HCO_3^- and the production of new HCO_3^-).

The physiological consequences of these changes to kidney function are profound and widespread. Whole body sodium retention occurs along with potassium depletion. As sodium retention occurs, water is also retained. Thus $[Na^+]_{plasma}$ does not change much. Plasma volume, however, is expanded and causes hypertension. The changes in electrolyte balance and pH alter the functions of excitable tissue such as nerve and muscle.

CLINICAL BACKGROUND

Primary hyperaldosteronism is a condition in which the adrenal cortex secretes aldosterone at a rate much greater than normal and in an uncontrolled manner. This condition usually results from the presence of an aldosterone-secreting adrenal adenoma (a usually benign neoplasm in which tumor cells form glandlike structures) in the adrenal cortex.

The symptoms that arise from this condition all derive from the excess mineralocorticoid activity of aldosterone: sodium and water retention, potassium depletion, and the expected consequences of expanded body fluid volume (hypertension) and Na/K imbalances (altered function of excitable membranes and muscles).

PROBLEM INTRODUCTION

Abnormal function in one organ or organ system rarely remains isolated to that system. This statement is particularly true when the abnormality lies in some component of the nervous or endocrine systems; these systems control the function of all the other organ systems. Disturbances to water and electrolyte balance, initiated by abnormal endocrine function and acting via altered function in the kidney, ultimately can cause cardiovascular problems as well.

Ms. R.C.V. is a 45-year-old Caucasian housewife. She has come to her physician complaining of extreme weakness. Her last visit was 5 years ago; at that time, her physician found that she had a mild hypertension and recommended that she restrict her salt intake.

Her physician took a history and did a physical examination, the results of which are provided later. Blood and urine samples were also taken for analysis. The results of these are given on the following page.

MEASUREMENT	PLASMA VALUE (normal range)	URINE VALUE
Na+	145 mEq/L	
K+	2.5 mEq/L	
HCO₃⁻	37 mM/L	0
Cl⁻	98 mEq/L	
pH	7.53	6.9
BUN	22 mg/dl (15-39)	
creatinine	0.71 mg/dl (0.53-1.19)	
protein	7.5 gm/dl (6.5-8.0)	negative

1. Which of the plasma and urine values are abnormal?

	PLASMA	URINE
Na+		
K+	LOW	
HCO₃⁻	HIGH	
Cl⁻		
pH	HIGH	LOW FOR PLASMA pH PRESENT
BUN		
creatinine		
protein		

2. What single pathology or pathophysiological condition could account for her abnormal plasma and urine values?

Critical facts are that plasma K is low whereas HCO₃⁻ and pH are high. The kidneys would normally compensate for an alkalosis by producing an alkaline urine. Ms. R.C.V.'s urine is too acid for the extracellular fluid pH that is present. It is important that attention be directed to all the data and any patterns that might be visible in that data. Possible causes include the following:

	RENAL	NONRENAL
LOW [K⁺]ₚ	EXCESS EXCRETION (LOSS)	Alkalosis resulting in shift of H⁺ out of cells and of K⁺ to intracellular compartment; excess loss of K⁺ from diarrhea or vomiting;
LOW [H⁺]ₚ	EXCESS EXCRETION (LOSS)	GI loss (vomiting); oral intake of base (antacid)
HIGH [HCO₃⁻]ₚ	EXCESS PRODUCTION	GI production with loss of H⁺ (vomiting); oral intake of base (antacid)

If the cause was renal or nonrenal in origin, you would expect to see the following:

	RENAL	**NONRENAL**
pH OF URINE	ACID TO WEAK ALKALINE	VERY ALKALINE

The data are consistent with the kidney being the source of the alkalosis; excess H^+ secretion causes an acid urine. The kidneys simultaneously produce excess HCO_3^-, thus increasing plasma concentration and causing the alkalosis. Non renal causes would be associated with renal compensation, with an alkaline urine.

The kidney-produced alkalosis could then be the cause of the hypokalemia. The problem, however, would remain: What could cause the kidneys to excrete excess acid? The usual cause of acid secretion by the kidneys is acidosis, but that condition is not present. Therefore, it is likely that the same stimulus that is causing the kidneys to increase acid secretion is causing the hypokalemia by causing increased renal K^+ excretion.

Thus the only situation that causes all plasma changes and generates an acid (or minimally alkaline) urine is when the kidney excretes excess K and excess H together. Other possibilities that cause alkalosis reactively yield very alkaline urine.

Both K and H are secreted in cortical collecting ducts; anything that caused this nephron segment to increase both K and H secretion could cause all of the findings. What could that be? The answer is aldosterone, because it stimulates secretion of both in this part of the nephron.

On the basis of the data available, the attending physician requested some special blood tests. They revealed that the patient's blood aldosterone level was *elevated* and that her plasma renin was *lower* than normal. The physician concluded that the patient most likely had an aldosterone hypersecreting adrenal tumor.

3. Predict the consequence of hyperaldosteronism on the following:

	RENAL EXCRETION	**PLASMA CONCENTRATION**
Na^+	DECREASED	NO CHANGE/INCREASE*
K^+	INCREASED	DECREASED
H^+	INCREASED	DECREASED
HCO_3^-	INCREASED	INCREASED

* Depends upon how much water is retained. [Na] is usually near normal, and is sometimes slightly elevated.

4. Draw a concept map that explains your predictions.

The increased aldosterone results in increased Na^+ reabsorption, which leads to an increased transluminal charge gradient that favors increased K^+ secretion and excretion. As K^+ is depleted from the renal tubule cells, H^+ secretion increases.

The increased Na reabsorption also results in increased water reabsorption. If water retention exactly matches Na retention, no change in $[Na]_{plasma}$ will result. If water retention is smaller than Na retention, there will be a small increase in $[Na]_{plasma}$.

The increased K excretion lowers $[K]_{plasma}$. In addition, the alkalosis that results causes K to enter the cells, further decreasing $[K]_{plasma}$. Figure renal 3-1 summarizes these changes.

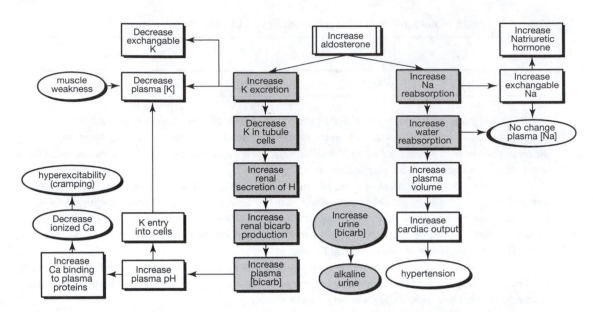

Figure renal 3-1. This concept map describes the alterations of function present in this patient as a consequence of abnormally high aldosterone levels. The highlighted section map illustrates the alterations of kidney function that are present.

The patient reports that she has gained some weight, which she attributes to getting older. She says that she has a tendency toward puffiness that she thinks is related to her menstrual cycle.

Physical examination revealed a woman with mild edema in her lower extremities. Her blood pressure was 150/95 and did not change after a period of rest. Ms. R.C.V. had complained about muscular weakness, indeed, her physician found her to be moving sluggishly, and her reflexes were distinctly slowed.

Her ECG was taken and revealed ST depression, a flattened T wave, and distinct U waves. A rhythm strip (a long-duration recording) showed occasional ventricular arrhythmias. All these electrocardiographic abnormalities are commonly seen with hypokalemia.

5. Identify the abnormalities in the patients history and physical.

CARDIOVASCULAR	NEUROMUSCULAR
WEIGHT GAIN	MOVING SLUGGISHLY
PUFFINESS	SLOW MOVING
HYPERTENSION	WEAKNESS
EDEMA	
ECG ABNORMALITIES	

6. Expand your concept map to explain these abnormalities, assuming that they all result from her endocrine condition.

The increased reabsorption of Na promotes reabsorption of water (which tends to follow Na movement osmotically). Plasma volume is increased with a consequent increase in central venous pressure, ventricular filling, and increased stroke volume and cardiac output. Hypertension will then be present. Figure renal 3-2 summarizes the origin of the patient's hypertension.

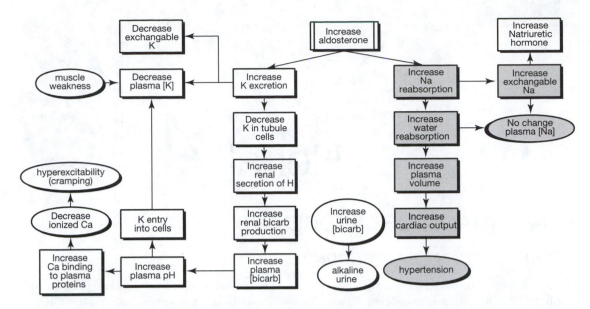

Figure renal 3-2. This concept map describes the alterations of function present in this patient as a consequence of abnormally high aldosterone levels. The highlighted section illustrates the origin of the patient's hypertension.

Aldosterone induces increased K secretion by the kidney, decreasing the body K content. It is difficult to predict what the membrane potentials of excitable cells might be because both intracellular and extracellular [K] have changed. (Alkalosis causes a shift of K into cells in exchange for H.) One could postulate that the neuromuscular changes seen in Ms R.C.V. result from changes in membrane potential, but this is not clear. Abnormal muscle structure is often seen with severe hypokalemia, and her neuromuscular symptoms may be a manifestation of structural damage to muscle cells.

Figure renal 3-3 describes the origin of the patient's neuromuscular problems.

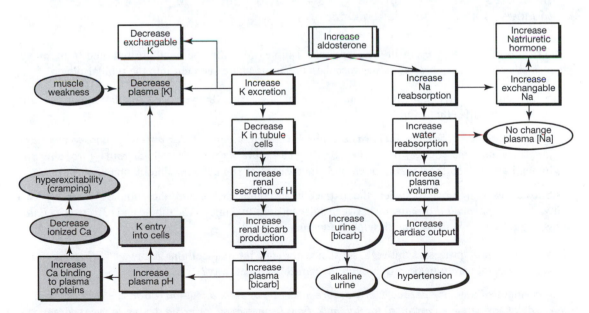

Figure renal 3-3. This concept map describes the alterations of function present in this patient as a consequence of abnormally high aldosterone levels. The highlighted section illustrates the origin of the patient's neuromuscular problems.

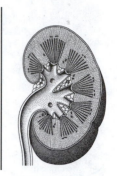

Renal 4

RENAL TRANSPORT PROCESSES

PHYSIOLOGICAL BACKGROUND

The transport of substances across cell membranes or across an epithelial cell layer occurs by a number of different mechanisms. Passive transport occurs when a substance diffuses down its electrochemical gradient. Primary active transport mechanisms directly use biological energy (ATP) to "pump" a substance against its electrochemical gradient. Secondary active transport occurs when the electrochemical gradient of one substance provides the energy for the transport of a second substance against its electrochemical gradient.

All these transport mechanisms are found in the nephron (and in the GI tract as well) and are involved in the kidney's regulation of body fluid composition.

PROBLEM INTRODUCTION

The kidneys play a major role in regulating the composition of body fluids. They carry out this function by transporting substances between the vascular compartment and the lumen of the nephron. An understanding of these transport processes and their controls is essential to understand the function of the kidneys as a homeostatic regulator.

Some of the substances present in the glomerular filtrate are reabsorbed and are returned to the body, whereas other substances present in the vascular compartment are secreted into the tubular lumen.

1. What routes are available for the movement of substances between the tubular lumen and the peritubular extracellular fluid?

   Substances can cross the wall of the nephron through a paracellular pathway (across the tight junctions present <u>between</u> the cells), or they can move via an intracellular route (crossing the luminal and basolateral membranes and the cytosol of the cells making up the nephron).

   Substances can move between the extracellular compartment and the tubular lumen by two pathways: (1) a *paracellular* pathway between adjacent cells making up the tubule and (2) a *transcellular* pathway through the walls of and through the tubular cells.

2. What transport processes move a substance through the paracellular pathway? What are some substances that are transported through the paracellular pathway?

   Movement of a solute paracellularly occurs either by passive diffusion (down the electrochemical gradient of that substance) or by bulk flow (sometimes referred to as solvent drag; the substance is carried in the water being osmotically transported). Substances transported via this pathway include water, Ca^{2+}, Mg^{2+}, and K^+.

3. What transport processes are involved in moving substances across the wall of the nephron by the intracellular pathway? What are some substances that are transported via the transcellular pathway?

Transport across cell membranes occurs by (1) passive diffusion, (2) facilitated diffusion, (3) primary active transport, or (4) secondary active transport.

Passive diffusion of a substance is driven by its electrochemical gradient. It usually occurs via ion channels in the luminal or basolateral membranes. It can also occur via the paracellular pathway. A small number of lipid soluble substances are able to diffuse through the lipid component of the membrane.

Facilitated diffusion occurs when a molecular carrier in the membrane is available to bind the substance and shuttle it across the membrane. This can involve simultaneous transport (cotransport) of two substances in the same direction (symport) or in opposite directions (countertransport or antiport). The transport, however, is passive in that it only occurs down an electrochemical gradient.

Primary active transport requires a membrane "pump" that directly uses ATP (biological energy) to move a substance against its electrochemical gradient. These "pumps" are, of course, proteins that span the cell membrane.

Secondary active transport occurs when two substances are cotransported and one of them is transported down its electrochemical gradient, providing energy for the other which is moved against its gradient. The substance moving against its electrochemical gradient is said to undergo secondary active transport.

Some substances transported via the transcellular pathway are Na, K, H, Cl, bicarbonate, and urea.

4. Give one example of a physiologically important substance that is reabsorbed or secreted by *primary active transport*. Describe the mechanism by which this substance is transported between the tubular fluid and the extracellular fluid.

A number of electrolytes are actively reabsorbed by the nephron, including Na^+, Ca^{+2}, and PO_4^{-2}. H^+ is actively secreted in several segments of the nephron.

The active transport of Na^+ is particularly important because the electrochemical gradient of this ion drives the secondary active transport of other substances such as glucose and amino acids.

The epithelial cells making up nephrons are asymmetrical, with the luminal and basolateral membranes having quite different properties. The membrane Na-pump (common to all cells) is located in the basolateral membrane. It maintains a low intracellular concentration of Na^+ by directly using ATP to "pump" Na^+ into the interstitium (while pumping K^+ into the cell). Na^+ enters the cell across the luminal membrane by passive diffusion through a Na channel and via facilitated diffusion with glucose, amino acids, etc. The net result is active reabsorption of Na^+ from the tubular lumen.

5. Give one example of a physiologically important substance that is reabsorbed or secreted by *secondary active transport*. Describe the mechanism by which this substance is transported. What determines the maximum rate at which this solute can be reabsorbed/secreted by this transport process? Of what significance is this limitation to the transport rate to the function of the kidney?

Glucose and amino acids are reabsorbed from the tubular lumen by a process of secondary active transport. Both enter the cell by facilitated diffusion, cotransported with Na^+. This process is driven by the concentration gradient for Na^+ that is established by the Na-pump. Because the cotransport protein does not directly use ATP, the movement of both glucose and amino acids is said to occur by <u>secondary</u> active transport.

The availability of carriers for Na + glucose and Na + amino acid determines the maximum rate at which reabsorption of glucose and amino acid can occur; when all the available carriers are occupied, transport will occur at the maximum rate. Thus reabsorption of glucose and amino acids both exhibit a transport maximum (T_m). If for whatever reason the filtered load of these substances exceeds the T_m, then the unreabsorbed part will be excreted.

6. Give one example of a substance that is reabsorbed or secreted passively. Describe the mechanism of this transport. What forces are present to drive this transport?

   Two substances of particular importance that are passively reabsorbed are water and urea. Water movement in biological systems is always passive and is driven by osmotic pressure gradients. Thus reabsorption of water is dependent on the existence of osmotic gradients created by the reabsorption (usually active) of solutes. Urea reabsorption depends on the reabsorption of water, which concentrates the urea and creates a concentration gradient for its movement.

7. Describe the handling of the following substances at each segment of the nephron by indicating for each segment whether net reabsorption (R) or net secretion (S) occurs there.

	NA+	K+	GLUCOSE	UREA	WATER
proximal tubule	R	R	R (COMPLETE)	R	R
descending limb of loop of Henle				S (TIP OF LOOP)	R
ascending limb of loop of Henle	R	R			
distal tubule	R	S			R
collecting duct	R	R AND S		R	R

8. Transport of some of these substances is subject to physiological controls via hormones that act on the kidney. In the table below, indicate which substances have transport processes that are controlled by hormones, where in the kidney this control is exerted, and what hormones are involved.

	NA+	K+	GLUCOSE	UREA	WATER
proximal tubule					
descending limb of loop of Henle					
ascending limb of loop of Henle					
distal tubule	ALDO	ALDO			
collecting duct	ALDO	ALDO		ADH	ADH

   Although plasma glucose concentration is normally well regulated, this process is not accomplished by the kidneys. Rather, it is the result of the controlled release of insulin, and other hormones, and the actions of these hormones in controlling cellular glucose uptake, utilization, and production.

9. List some other substances whose reabsorption in the kidney is under endocrine control.

   The reabsorption of calcium and phosphate in the kidney is also regulated by the endocrine system. Parathyroid hormone increases the tubular reabsorption of calcium, whereas the tubular reabsorption of phosphate is inhibited by parathyroid hormone but is stimulated by vitamin D_3.

Renal 5

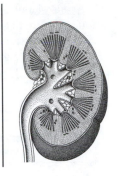

ACTION OF DIURETICS

PHYSIOLOGICAL BACKGROUND

For the kidneys to produce a urine that is more concentrated or more dilute than body fluids requires that the nephron be able to reabsorb more or less water than the amount of solute it reabsorbs at the same time. Water movement, however, is always passive and is driven by an osmotic pressure gradient. As a consequence, if the tubule is permeable to water, reabsorption of solute will always be accompanied by reabsorption of enough water to create an isosmotic (to plasma) fluid.

The mammalian kidney is able to produce a urine the concentration of which can be varied to maintain the osmolarity of body fluids. The kidney accomplishes this in the following manner.

The ascending limb of the loop of Henle actively pumps NaCl out of the lumen and into the interstitial space. Altough the pump is only able to produce an osmotic gradient of approximately 200 mOsm/L at any point, the "hair pin" geometry of the nephron creates a countercurrent multiplier that results in an osmotic gradient: 300 mOsm/L in the cortex to 1200 mOsm/L in the papilla. Simultaneously, the tubular fluid is diluted. As the dilute tubular fluid enters the collecting duct and flows from the cortex to the medulla, it is exposed to this gradient of increasing osmolarity. If the water permeability of the tubular membrane is high (which is the case if the circulating concentration of ADH is high), then water will be reabsorbed and a concentrated urine will result. If ADH concentration is low, then little or no additional water is reabsorbed and a dilute urine is produced. The secretion of ADH is a function of hypothalamic osmolarity, thus establishing a homeostatic mechanism to regulate body fluid osmolarity, with the kidney acting as the major effector in the reflex loop.

Alterations to any aspect of this renal concentrating/diluting mechanism through administration of pharmacologic agents can give rise to increased urine flow rate or diuresis.

CLINICAL BACKGROUND

Diuretics are a commonly prescribed class of drugs. They increase the rate of urine production (the urine flow rate) by reducing the reabsorption of solute (principally NaCl) at different segments of the nephron. Thus not only is water excreted at a faster rate, but solute is also lost from the body.

These drugs are used to treat a number of diseases including hypertension, congestive heart failure, renal disease, and any condition in which generalized edema is present.

Diuretics have a variety of effects in addition to increasing urine flow rate and solute loss, and these side effects can pose serious consequences for body homeostasis.

PROBLEM INTRODUCTION

The kidneys have the ability to vary the osmolarity of the urine that they produce, which enables them to regulate body fluid osmolarity. The renal mechanisms involved are complex and involve every segment of the nephron. Disturbances to any of these functions will result in significant changes in the urine produced and in the homeostatic state of the body.

1. <u>Acetazolamide</u> is a drug that inhibits the enzyme carbonic anhydrase in the cells of the proximal tubule. Predict the effects (increase/decrease/no change) of acetazolamide on the following parameters (consider only the direct effects of the drug on the kidney and ignore any reflex responses that might be generated).

	URINE CONCENTRATION OF				URINE FLOW RATE
	Na$^+$	K$^+$	Cl$^-$	HCO$_3^-$	
acetazolamide	INCREASE	INCREASE	DECREASE	INCREASE	INCREASE

2. Explain the mechanism by which acetazolamide produces these effects.

Acetazolamide enters the proximal tubule cells, where it inhibits the action of carbonic anhydrase (CA). This enzyme is essential for the rapid hydration of CO_2 to form H_2CO_3, which subsequently dissociates into H^+ and HCO_3^-. Thus, with CA inhibited, the availability of protons for Na^+-H^+ exchange across the luminal membrane is reduced and Na^+ reabsorption is greatly reduced. At the same time, bicarbonate reabsorption is reduced because less H^+ is available to react with HCO_3^- in the tubular lumen. Normally water passively follows the reabsorbed solute. With less Na^+ and HCO_3^- reabsorbed, there is less water reabsorbed, which reduces the concentration of K^+ in the lumen and reduces the passive K^+ reabsorption. Cl^- concentration in the tubular fluid is reduced by the increased volume of water remaining there, and the smaller net cation loss from the tubular fluid reduces the potential gradient favoring reabsorption.

3. <u>Furosemide</u> is a drug that inhibits the active reabsorption of solute in the ascending limb of the loop of Henle by blocking the Na, K, 2Cl co-transporter in the luminal membrane. Predict (increase/decrease/no change) the effects of furosemide on the following parameters (consider only the direct effect of the drug on the kidney and ignore any reflex responses that might be generated).

	URINE CONCENTRATION OF				URINE FLOW RATE
	Na$^+$	K$^+$	Cl$^-$	HCO$_3^-$	
acetazolamide	INCREASE	INCREASE	DECREASE	INCREASE	INCREASE
furosemide	INCREASE	INCREASE	INCREASE	INCREASE	INCREASE

4. Explain the mechanism by which furosemide produces these effects.

Approximately 25% of the filtered NaCl is actively reabsorbed in the thick ascending limb of the loop of Henle. Furosemide inhibits the Na, K, 2Cl symporter in the luminal membrane and thus inhibits the reabsorption of solute, with two consequences. First, more Na, Cl, and K are lost in the urine. At the same time, the solute concentration in the renal interstitium is reduced, which in turn reduces the interstitial osmotic concentration and reduces the reabsorption of water from the collecting duct. A marked diuresis (increased urine flow) results. The increased flow of fluid into the distal tubule and collecting duct reduces the concentration of K in the lumen and results in a steeper gradient for K diffusion out of the cells into the lumen; hence, K in the urine increases further.

5. Thiazide is a drug that decreases the reabsorption of Na in the early portion of the distal tubule. Predict the effects of thiazide (increase/decrease/no change) on each of the following parameters.

	URINE CONCENTRATION OF				URINE FLOW RATE
	Na^+	K^+	Cl^-	HCO_3^-	
acetazolamide	INCREASE	INCREASE	DECREASE	INCREASE	INCREASE
furosemide	INCREASE	INCREASE	INCREASE	INCREASE	INCREASE
thiazide	INCREASE	INCREASE	INCREASE	INCREASE	INCREASE

6. Explain the mechanism by which thiazide produces these effects.

Thiazide inhibits the entry of Na into the cells in the early distal tubule (by a mechanism that has not yet been characterized). As a consequence, Na reabsorption is reduced (as is the reabsorption of Cl that normally follows Na down its electrochemical gradient). The reduced Na reabsorption reduces water reabsorption, and a diuresis results. There is thus increased flow into the late distal nephron and collecting duct. The concentration of K in lumen of the later distal tubule and collecting duct is reduced, favoring diffusion of K into the lumen (K secretion and excretion). The increased loss of K^+ from the cells results in reduced secretion of H^+ and hence reduced reabsorption of HCO_3^-.

7. Spironolone is a drug that blocks the action of aldosterone on the collecting duct. Predict the effects (increase/decrease/no change) of spironolone on the following parameters.

	URINE CONCENTRATION OF				URINE FLOW RATE
	Na^+	K^+	Cl^-	HCO_3^-	
acetazolamide	INCREASE	INCREASE	DECREASE	INCREASE	INCREASE
furosemide	INCREASE	INCREASE	INCREASE	INCREASE	INCREASE
thiazide	INCREASE	INCREASE	INCREASE	INCREASE	INCREASE
spironolone	INCREASE	DECREASE	INCREASE	INCREASE	INCREASE

8. Explain the mechanism by which spironolone produces these effects.

Aldosterone acts on the cells of the late distal tubule and the collecting duct to increase the reabsorption of Na. It does this through a number of intracellular changes. The number of channels in the luminal membrane through which Na can enter the cell is increased, allowing a greater flux of Na into the cell (down its electrochemical gradient). At the same time, the activity of the basolateral Na/K- ATPase is increased, resulting in a greater rate of pumping Na into the interstitium. This increase results, at least in part, from changes to cell metabolism that make available greater amounts of ATP to power the pump.

Spironolone is a competitive inhibitor of aldosterone (it binds to the same receptor sites). Hence, reabsorption of Na is reduced (excretion of Na^+ increases) and the secretion of K^+ is reduced (excretion of K^+ is reduced). The increase in K^+ in the tubule cells results in less H^+ being secreted into the lumen and therefore less HCO_3^- is reabsorbed (bicarbonate excretion is increased).

9. Triamterene is a drug that blocks Na reabsorption in the collecting duct by blocking the pumping of Na into the cell. Predict the effects of triamterene (increase/decrease/no change) on the following parameters.

	URINE CONCENTRATION OF				URINE FLOW RATE
	Na^+	K^+	Cl^-	HCO_3^-	
acetazolamide	INCREASE	INCREASE	DECREASE	INCREASE	INCREASE
furosemide	INCREASE	INCREASE	INCREASE	INCREASE	INCREASE
thiazide	INCREASE	INCREASE	INCREASE	INCREASE	INCREASE
spironolone	INCREASE	DECREASE	INCREASE	INCREASE	INCREASE
triamterene	INCREASE	DECREASE	INCREASE	INCREASE	INCREASE

10. Explain the mechanism by which triamterene produces these effects.

Triamterene produces a diuresis that is essentially identical to that produced by spironolone. But, triamterene does not compete with aldosterone, and thus it is effective even in the absence of aldosterone. With reduced Na reabsorption, less Cl will follow. The increased concentration of Na in the tubular fluid results in reduced excretion of K and H. With less H being secreted, less bicarbonate is reabsorbed.

11. Diuretics are not uncommonly self-prescribed to achieve weight loss for cosmetic or sports-related reasons. What homeostatic imbalances might result from such medically uncontrolled use of these drugs?

The three principal dangers of abuse (excess use) of diuretics are dehydration, disturbances of potassium balance, and acid-base disturbances. All diuretics promote loss of water from the body. Many diuretics (acetazolamide, furosemide, thiazide) also promote loss of potassium and can lead to hypokalemia unless countered by simultaneous increase in the dietary intake of potassium. On the other hand, spironolone and triamterene reduce the loss of potassium and can lead to hyperkalemia. The role of potassium in establishing the resting membrane potential and hence the excitability of nerves and muscle makes disturbances to potassium balance (in either direction) dangerous. Acid-base disturbances have similar widespread effects because the function of all enzymes is pH-dependent.

12. What is the role of active transport in enabling the kidneys to produce a urine that is more concentrated, or less concentrated, than body fluids?

To produce a concentration urine, one that has a higher osmolarity than plasma, the kidney must reabsorb more water than solute. But because water movement is always passive, down an osmotic gradient, this would seem to be impossible to achieve.

    In the mammalian kidney, the shape of the nephron, the appropriate localization of active pumps that transport NaCl from tubular lumen to renal interstitium, and the water impermeability of the ascending limb of Henle's loop give rise to a phenomenon called countercurrent multiplication. Although the pumps in the ascending limb of the loop of Henle can only establish an osmotic gradient of approximately 200 mOsm/L across the wall of the tubule, the counter-current multiplier creates an interstitial osmotic gradient of 300 mOsm/L in the cortex to 1200 mOsm/L in the renal papilla. Thus, as tubular fluid passes down the collecting duct from the cortex to the papilla, it is exposed to an increasing osmotic gradient. If the permeability of the collecting duct is high (if the concentration of ADH is high), water will be reabsorbed without additional solute being removed from the fluid. A concentrated urine results.

UNIT 8

Endocrine 1

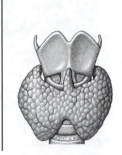

ENDOCRINE CONTROL SYSTEMS

PHYSIOLOGICAL BACKGROUND

Endocrine glands are part of homeostatic control systems. There are two large classes of such systems: (1) those in which hormone secretion is determined by the plasma concentration of a regulated nonhormone material and (2) those in which hormone secretion is determined by the plasma concentration of a regulated hormone. In the first instance, the primary stimulus for hormone secretion is a change in the plasma concentration of a substance whose concentration the endocrine system regulates. In that case, the gland acts as both a sensor for the regulated variable and a controller of the effectors that carry out the regulation. For example, the parathyroid (PT) gland regulates the plasma concentration of calcium ($[Ca^{2+}]_p$), and the primary stimulus for PT hormone secretion is a decrease in plasma $[Ca^{2+}]$. The PT senses the change in $[Ca^{2+}]_p$ and controls the effectors that determine the value of $[Ca^{2+}]_p$.

In contrast, plasma thyroid hormone (TH) concentration is itself regulated. The secretion of TH is determined by the plasma concentration of thyroid stimulating hormone (TSH), and TH controls the secretion of TSH by negative feedback on the anterior pituitary gland and possibly the hypothalamus. In such a system, the metabolic actions of TH play no significant role in determining TH secretion rate.

PROBLEM INTRODUCTION

All the endocrine systems are organized in a manner that permits them to carry out their function: the control of bodily activity in the service of homeostasis. Such systems respond to disturbances so as to minimize the effect of these disturbances on the system's regulated variables.

Figure endo 1-1, p. 330 is a diagram of a hypothetical endocrine control system. The following abbreviations are used here and in the figure:

$$H = \text{the hormone product of the endocrine cell}$$

$$C = \text{a specific plasma carrier of hormone H}$$

$$X = \text{a normal stimulus to the endocrine cell}$$

The lowercase letters in the figure refer to situations defined below. Determine the consequences of each situation on the hormone concentration in the blood ([H]) and the hormone EFFECT. As used here, EFFECT is a gross action of a hormone. For example, an EFFECT of insulin is to reduce the plasma

329

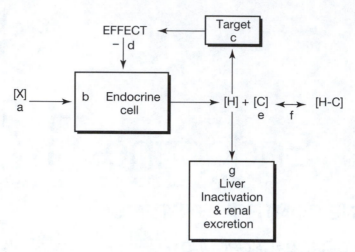

Figure endo1-1. A hypothetical endocrine control system that regulates the plasma concentration of the hormone H ([H]). The lowercase letters refer to various situations that act as disturbances to the system. See the text for a description of the situations.

glucose concentration. Assume that the condition described in each situation is maintained for a long time. Predict the changes you expect to occur in each situation in two time periods: (1) the immediate consequences and (2) the steady-state consequences. Use $++$ for a large increase, $+$ for a smaller one; $--$ for a large decrease, $-$ for a smaller one; and 0 for no change or a return to the initial condition.

SITUATION	IMMEDIATE CONSEQUENCES		STEADY-STATE	
	[H]	EFFECT	[H]	EFFECT
a	$++$	$++$	$+$	$+$
b	$--$	$--$	$0/-$	$0/-$
c	$++$	$--$	$+$	$0/-$
d	$--$	$--$	$-$	$-$
e	$--$	$--$	0	0
f	$++$	$++$	0	0
g	$++$	$++$	$0/+$	$0/+$

SITUATIONS:

a. [X] is increased. Draw a graph of [H] versus time starting before the increase in [X] and continuing into the new steady state.

b. The biochemical pathways that enable X to stimulate H secretion are partially inhibited.

c. The number of H receptors is decreased.

d. The endocrine cell response to the EFFECT is facilitated.

e. The plasma carrier concentration, [C], is suddenly increased. Draw a graph of [H] and [HC] on the same time axis starting before the increase in [C] and continuing to the new steady state.

f. The hormone binding constant of C is decreased.

g. Liver or renal failure leads to a decrease in H inactivation or excretion.

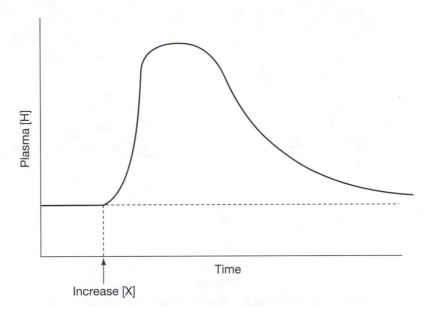

Figure endo 1-2. An increase in [X] causes an initial rise in plasma [H]. Feedback inhibition, however, restores the value of this regulated variable toward normal. In the final steady state, plasma [H] is only slightly elevated. See the text for a detailed description of the response.

Explanations are given below.

a. An increase in [X] immediately stimulates the endocrine cell to secrete more H. Hence, plasma [H] increases. The increased [H] stimulates the target more, causing the EFFECT to increase. The increased EFFECT then inhibits endocrine cell H secretion (negative feedback). That in turn causes plasma [H] to fall, lowering the EFFECT from its previous high. Neither [H] nor the EFFECT, however, can return to normal. If they did, the original disturbance (increased [X]) would again be acting unrestrained. In the steady state, a sustained increase in [H] is present and its feedback effects offset most of the consequences of the maintained increase in [X]. See Figure endo 1-2.

b. Inhibiting the effect of X on the synthesis of H will immediately cause plasma [H] to decrease, which reduces the stimulation of the target. Thus the EFFECT will fall, decreasing the inhibition of the endocrine cell. As a result, H secretion will increase, raising [H] and the EFFECT. In the steady state, both [H] and the EFFECT could be returned to normal. If the stimulating action of X is an overwhelmingly essential EFFECT, however, both will remain reduced.

c. With fewer target cell receptors, the existing [H] will produce a smaller EFFECT, removing part of the feedback inhibition of the endocrine cell. H output and [H] will therefore increase. If the loss of receptors is not too severe, the elevated [H] will restore the EFFECT to normal in steady state. If the loss of receptors is severe, the EFFECT will remain reduced.

d. If the endocrine cell response increases, the same EFFECT will cause a greater feedback inhibition and will decrease H secretion. The consequent fall in [H] will stimulate the target less, causing the EFFECT to decrease. Thus the feedback inhibition will be reduced, allowing both [H] and the EFFECT to partially return to their normal values in steady state.

e. A sudden increase in [C] will tie up some of the free H, causing [H] to fall. The target will be stimulated less and will reduce the EFFECT. Thus the feedback inhibition of the endocrine cell will be reduced, increasing H secretion and [H]. In the steady state, total H (free H + H-C) will be larger than before [C] was increased, but [H] will be returned to normal. Hence the EFFECT will

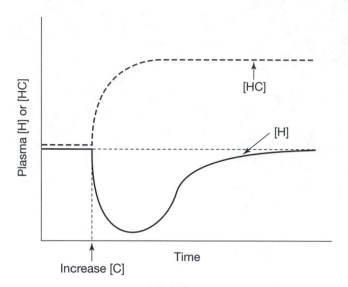

Figure endo 1-3.   Increasing the plasma hormone carrier concentration ([C]) ties up free hormone, decreasing the plasma concentration of free hormone ([H]). This action decreases the target stimulation, causing a decreased EFFECT and less feedback inhibition of the endocrine cell. H secretion will therefore rise, replacing the lost free plasma [H]. In the steady state, [H] is restored to normal but [HC] remains higher than normal. See the text for a further description.

also be restored. The excess H secretion "saturates" the added carrier, but once this happens, the original H secretion rate is sufficient to maintain the desired (set point) plasma [H]. See Figure endo 1-3.

f.   A <u>decrease in the binding of H by C</u> is the reverse of the situation in situation e. The immediate result is liberation of H from C. Both [H] and then the EFFECT will increase. In the steady state, however, both will be completely restored by the feedback system as the increased EFFECT decreases H secretion and [H]. Total [H]- that is, H + H-C—will be lower than it was originally, however.

g.   <u>Reduced inactivation or excretion of H</u> allows [H] to increase, stimulating the target and causing EFFECT to rise. This action increases the inhibition of the endocrine cell, decreases its H secretion, and lowers [H] and the EFFECT in turn. In the steady state, if H secretion can be reduced sufficiently, [H] and the EFFECT will be restored to normal. If it cannot, both will remain somewhat elevated.

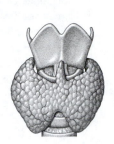

Endocrine 2

ACROMEGALY AND HYPOPITUITARISM

PHYSIOLOGICAL BACKGROUND

The pituitary gland is anatomically situated in the sella turcica, a cavity in the sphenoid bone. It is connected via the pituitary stalk to the hypothalamus. Long neural tracts traverse the stalk and connect the hypothalamus to the posterior pituitary, and the hypothalamic-pituitary portal system, a circulatory pathway that delivers hypothalamic hormones, connects the hypothalamus directly to the anterior pituitary (see Figure endo 2-1).

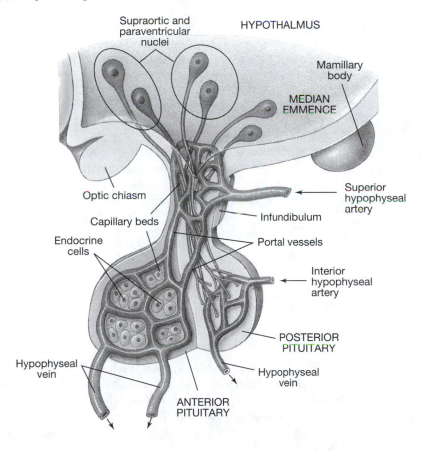

Figure endo 2-1. A diagram of the anatomy of the pituitary gland and its relationship to the hypothalamus and the optic chiasm. Martini, *Fundamentals of Anatomy and Physiology*, 4th ed. Prentice Hall, 1998. p. 604

Histologically, the anterior pituitary is composed of gland cells, each of which secretes specific hormones and has characteristic staining that is determined by the histochemical characteristics of the hormones it secretes. For example, growth hormone–(GH, also called somatotropin, STH) secreting cells stain with the acid dye, eosin, and are classified as acidophilic or eosinophilic.

Hormone secretion is a carefully controlled process. Each anterior pituitary hormone is secreted at a rate that is determined by: (1) the effects of its respective hypothalamic releasing and/or inhibiting hormone(s) and (2) inhibitory feedback by target gland hormones. Other influences, some of which act through the hypothalamus and some on the target glands' cells, have stimulatory and inhibitory effects. The result is the regulation of the plasma hormone concentration or the homeostatic regulation of functions that are controlled by each of the hormones.

CLINICAL BACKGROUND

A significant disturbance applied to an endocrine system will alter its hormone secretion rate and cause characteristic signs and symptoms of hormone excess or deficiency. A common cause for changes in hormone secretion is the development of a tumor composed of endocrine cells. Such tumor cells change or loose their responsiveness to normal feedback controls and become hypersecreting.

Tumor cells also lose the growth restraints that keep them under check, and they multiply excessively and may impinge on the activity of adjacent cells. This condition is particularly serious when the tumor is growing in a confined space. The site of the pituitary gland, the sella turcica, is such a confined space. Hyperplasia of one type of pituitary cell may compress other cells, reducing their ability to function or killing them.

When pituitary tumors grow large enough, the entire gland may bulge superiorly and compress the optic chiasm, blocking conduction in the crossing fibers. These fibers come from the nasal parts of the retinae and serve the lateral visual fields. The tumor may bulge laterally and compress cranial nerves passing through the cavernous sinuses. These nerve fibers innervate the extraocular muscles and the autonomic supply to the pupils.

PROBLEM INTRODUCTION

The anterior pituitary gland secretes many hormones, some of which are trophic to peripheral endocrine glands. Changes in the secretion rate of any of these hormones produce characteristic effects. This problem deals with a situation in which a patient experiences changes in hormone secretion.

Medical history

A 51-year-old man came to the emergency room. He complained of 4 days of a retrobulbar headache of increasing severity. He also reported 8 years of decreased libido, 3 years of chronic fatigue, cold intolerance, decreased beard growth, and poor tanning of the skin. He has had a progressive increase in shoe and ring size over the past 20 years, along with increasing enlargement of his nose, lips, tongue, and jaw.

Physical examination

The patient's blood pressure (BP) was 100/60 mm Hg lying and 80/50 mm Hg standing. His skin was dry and pale, and it appeared thickened. Pubic and axillary hair were scanty. His testes were smaller and softer than normal. Deep tendon reflexes showed a prolonged relaxation phase. His nose, jaw, zygomata, and lips were prominent; his tongue was large, and the soft tissue of his hands and feet were enlarged. He exhibited bitemporal hemianopsia (blindness in his lateral visual fields); his left pupil was dilated, and his left eye deviated laterally.

1. In the table below, list the findings in the patient's history and physical that seem to be abnormal. Collect these into groups, each of which seems to you to be associated with a single hormone system. Identify the hormone associated with each group and indicate whether the data suggest an increase or a decrease (I or D) in hormone secretion rate.

HORMONE	I/D	ABNORMALITIES
Male sex hormone (testosterone)	D	decreased libido
		decreased beard growth
		scanty pubic and axillary hair
LH and FSH	D	testicular atrophy
Thyroid hormone	D	chronic fatigue
		skin dry and pale
		tendon reflex has slow relaxation
		soft tissue growth (myxedema)
		cold intolerance
Growth hormone	I	Increased shoe and ring size
		enlarged nose, lips, tongue, and jaw
ACTH or MSH	D	poor skin tanning
Cortisol	D	orthostatic hypotension (excessive fall in BP upon standing)

Initial Lab Workup

Laboratory studies were ordered with the following results.

TEST	RESULTS
urinalysis	unremarkable
white blood count	6000 with 6% eosinophils (normal <3%)
serum sodium	128 mEq/L
serum potassium	4.7 mEq/L
plasma cortisol	3µg/dl (normal 7–25 at 8:00 A.M.)

2. Do these results confirm or deny any of your hypothesized hormone imbalances? Explain how you used the data to confirm or deny your hypotheses.

Decreased plasma cortisol confirms hypoadrenalcorticalism. Cortisol is lympholytic and eosinophilolytic. Decreased cortisol is the probable cause of the increase in circulating eosinophils.

Decreased [Na] could result from two causes. First, low cortisol secretion prevents the kidneys from excreting H_2O at a normal rate (mechanism unknown), which dilutes the plasma Na. Second, cortisol has mineralocorticoid actions (not as much as the mineralocorticoids do, but cortisol is normally present in very much higher concentration than are the mineralocorticoids). A large reduction in plasma cortisol concentration would decrease renal Na reabsorption and reduce the plasma [Na]. It would also decrease renal K secretion and account for the elevated (high normal) plasma [K].

Over the next 8 hours, the patient was given fluids and hydrocortisone intravenously. His supine BP increased to 120/80 mm Hg, his headache worsened, and his condition obviously deteriorated.

3. Explain what happened to the patient during this period.

Insufficient cortisol reduces the vascular response to catecholamines, which would cause orthostatic hypotension (less reflex vasoconstriction upon arising) and contribute to his low BP. Hydrocortisone replacement would help to restore his vascular responsiveness to ANS inputs and raise his BP. The patient received fluids, which expanded his blood volume and also contributed to his increased BP. The increased BP increased the cranial tissue volume and pressure, worsening his headache.

Endocrine Lab Workup

The results of other laboratory studies have just been received. The specimens were drawn on admission, but results have just been returned.

TEST	RESULTS
serum thyroxine	3.1 µg/dl (normal 4–11)
serum TSH	<1µU/ml (normal 0–10)
serum testosterone	189 ng/100 ml (normal male 300–1100)
serum LH	1.7 mIU/ml (normal 1–15)
serum PRL	4 ng/ml (normal male, <15)
serum GH	47 ng/ml (normal <5)

4. What is your final diagnosis? Explain your reasoning.

There is hypersecretion of growth hormone along with hyposecretion of ACTH, TSH, LH, and PRL. Hyposecretion of the trophic hormones caused a decrease in the secretion of cortisol, thyroid hormone, and testosterone.

The gonadal-related and thyroid-related symptoms are due to hypopituitarism and not to primary failure of the peripheral glands. A primary deficiency of testosterone and thyroid hormone would reduce feedback inhibition of the hypothalamus and pituitary. As a consequence, plasma LH and TSH would be elevated. The plasma concentrations of LH and TSH are both low, however.

The patient's problem could not be due to a general reduction in hypothalamic secretion. The hypothalamus has an inhibitory effect on PRL secretion. Hence, a reduction in hypothalamic secretion would cause PRL concentration to be higher than normal. Yet, plasma [PRL] is low. Hence, one may also conclude that the reduction in plasma cortisol concentration resulted from reduced ACTH secretion by the pituitary itself, not from reduced CRH secretion by the hypothalamus.

The patient has had the symptoms of acromegaly (excess GH secretion) for at least 20 years, and the symptoms of reduced pituitary secretion of the other hormones has been present for 8 years. The most likely explanation is that he has a hypersecreting pituitary somatotrope tumor, which compressed the remaining anterior pituitary cells, reducing their secretion. The compression of other intracranial structures is probably the cause of the patient's headache. When his BP was raised, enlarging the size of the pituitary, his headache got worse. The increased BP may also have caused pituitary apoplexy (hemorrhaging into the tumor), a new, acute problem.

The presence of the tumor also explains his bitemporal hemianopsia. As the tumor enlarged it compressed his optic chiasma, which blocked the impulse conduction in the nerve fibers supplying the nasal portion of both retinas (they provide vision in the temporal visual fields). Compression of the cranial nerves in the cavernous sinuses is the probable cause of the loss of pupillary control and the lateral deviation of his left eye.

The patient is suffering from a growth hormone–secreting tumor of his anterior pituitary, acromegaly, of 20 years duration. In addition, several features of hypopituitarism have been present for at least 8 years.

5. Draw a concept map relating the patient's pathology to his signs and symptoms.

   See Figure endo 2-2.

6. How would you treat this patient? What long-term consequences would you have to be prepared to deal with and how would you handle them?

   The tumor would normally be removed surgically, by performing a hypophysectomy. This procedure requires that replacement therapy of the hormone essential to life (cortisol) be maintained. Therapy with hormones that are not required to sustain life but that ensure an acceptable quality of life would also be prescribed. When the patient recovers from the surgery, replacement therapy with L-thyroxine and testosterone will be started.

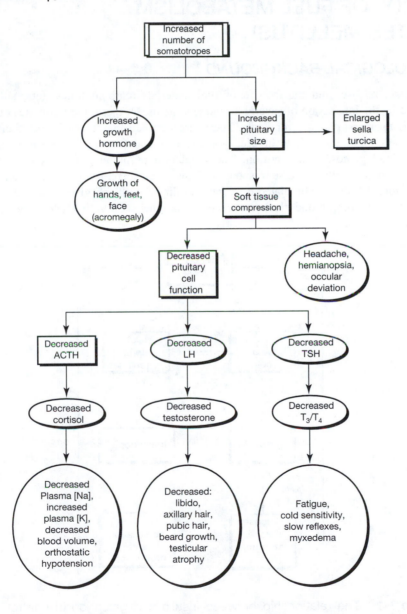

Figure endo 2-2. The causal relationships that explain the effects of a hypersecreting anterior pituitary tumor made up of somatotrophic cells.

Endocrine 3

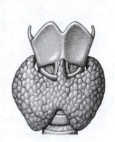

CONTROL OF FUEL METABOLISM (DIABETES MELLITUS)

PHYSIOLOGICAL BACKGROUND

The storage, release, and metabolism of fuel materials (carbohydrates, fats, and proteins) are controlled by the interaction of several hormone systems. Of these, the hormones of the pancreatic islets of Langerhans play a central role. The interaction between the fuel-storing effects of insulin and the fuel-mobilizing effects of glucagon are essential for the maintenance of a reasonably constant blood glucose concentration. The regulation of blood glucose concentration is made possible in part by the inverse response of the islet secretory cells to changes in glucose concentration: β cells are stimulated to secrete insulin when blood glucose concentration rises and are inhibited by falling blood glucose concentration, whereas α cells react in the opposite manner (see Figure endo 3-1).

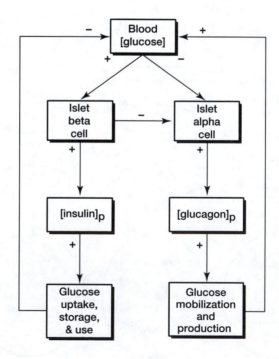

Figure endo 3-1. The relationship between insulin and glucagon in controlling the blood glucose level.

Insulin and glucagon also affect the metabolism of fats. Insulin stimulates lipogenesis and inhibits ketogenesis. Glucagon does the opposite. Finally, insulin stimulates amino acid uptake and protein synthesis.

Other, so-called counterregulatory hormones (epinephrine, growth hormone, cortisol) mostly oppose the actions of insulin on carbohydrate, fat, and protein metabolism.

CLINICAL BACKGROUND

Diabetes mellitus is the most common endocrine disorder known. Two forms of the disease exist: type I or juvenile-onset diabetes and type II or adult-onset diabetes. In type I diabetes, there is a severe reduction, or a total absence, of insulin secretion. Fasting blood glucose is elevated. In type II diabetes, target cells exhibit a reduced sensitivity to insulin. Insulin secretion is actually greater than normal in some type II diabetics, although, it is lower than normal in others. Fasting blood glucose may be normal or elevated, but glucose intolerance is always present (i.e., a standard glucose load produces a larger than normal increase in blood glucose concentration, and the increase lasts longer than normal).

Type I diabetics have insulin-dependent diabetes mellitus (IDDM). If a type I diabetic is not given insulin, all the actions of insulin are reduced or absent and the insulin-opposing effects of the counterregulatory hormones act unimpeded. As a consequence, plasma glucose concentration is high, which increases the extracellular fluid osmolarity and produces the osmotic diuresis that gives the condition its name (diabetes means increased urine flow). Lipid mobilization and ketogenesis are also present in untreated type I diabetics.

In type II diabetics, however, the plasma insulin concentration may be high enough to oppose many of the anti-insulin effects of the counterregulatory hormones. Untreated type II diabetics do not usually exhibit ketogenesis, but they may become hyperosmolar.

Ketones (β-hydroxybutyric acid and acetoacetic acid) are metabolic products of long-chain fatty acids. When produced in moderate amount, ketones are excellent fuel materials. When produced in excess (e.g., in uncontrolled type I diabetics), however, they cause acidosis, a condition that alters many body functions.

PROBLEM INTRODUCTION

Most hormones have multiple direct effects in the body. These actions influence many more body systems, in turn, so an imbalance in a single system often has widespread consequences. This problem illustrates such a situation.

A mother brought her 13-year-old daughter, Jenny, to her family physician. The mother was concerned about what she described as "emotional disturbances" in the child (bed-wetting and generalized "laziness"). Upon questioning, the daughter revealed that she had recently developed a habit of eating between meals and drinking a great deal of water to satisfy her thirst. She had started having menstrual periods 6 months earlier.

1. Define two possible pathophysiological mechanisms that could give rise to the symptoms described in this history.

   The patient has increased food intake (polyphagia), thirst, increased drinking (polydipsia), bed-wetting, and laziness.

   a. *Diabetes mellitus: insufficient insulin. This condition would explain the increased eating and drinking, cardinal signs of the disease. The osmotic diuresis (polyuria) seen in diabetes mellitus increases her urinary volume and could be the cause of her bed-wetting. Her laziness could result from insufficient fuel supply to her insulin-dependent tissues.*

   b. *Diabetes insipidus: insufficient ADH. This condition does not explain all of the symptoms, but it does explain the bed-wetting (polyuria) and excess drinking (polydipsia).*

The patient appeared to be a rather thin but otherwise normal adolescent. She did, however, have glucosuria (sugar in the urine) and a blood sugar of 350 mg/dl.

2. What disease does this patient have?

> **Diabetes mellitus. ADH deficiency is ruled out because diabetes insipidus does not cause hyperglycemia.**

The patient is diabetic. Her blood sugar was brought under control with daily insulin injections and has been well controlled since then.

At age 16 (3 years after her first hospitalization), the patient was admitted to the emergency room in a coma. The mother reported that her daughter had attended a school dance the previous evening and had arrived home very late. She was allowed to sleep the next morning, and no attempt had been made to wake her until well after noon. At that time, she could not be adequately aroused.

Examination revealed a comatose patient with slow but very deep respirations. Her urine was positive for glucose and ketone bodies.

3. Why has the patient become ill?

> **She did not take her insulin. As a result, her fuel metabolism became uncontrolled.**

Jenny's blood chemistries from a sample drawn on admission to the ER were the following:

BLOOD CHEMISTRY	VALUE
Na$^+$	152 mM
K$^+$	7.1 mM
Cl$^-$	88 mM
HCO$_3^-$	10.8 mM
pH	7.27
BUN	35 mg/dl

4. Explain each of the patient's signs, symptoms, and lab values while she was comatose.

> **With inadequate plasma insulin, she has increased glucose mobilization, increased glucose production via glycogenolysis and gluconeogenesis, and decreased glucose utilization. Therefore, her plasma glucose is elevated. The primary substrates for gluconeogenesis are amino acids, which are deaminated and the amino groups converted to urea. This causes the increase in her blood urea nitrogen (BUN), a measure of blood urea concentration. See Figure endo 3-1.**
>
> **She became comatose for the following reasons. The increase in her plasma glucose and ketone concentrations increased her plasma osmotic concentration, dehydrating her cells and concentrating their contents. This process caused an increased [K] in the neurons, tending to increase their resting membrane potential, which would make it harder to excite the cells. Increased urinary water loss (in excess of electrolyte loss) would, however, tend to concentrate extracellular Na$^+$ and K$^+$. Hence, although the decrease in CNS excitability is clear, the actual change in resting potential is uncertain. See Figure endo 3-2.**
>
> **She is mobilizing fatty acid from her triglyceride stores, and her hepatic ketone production is greatly increased because of reduced inhibition by insulin and increased stimulation by glucagon. Both glucose and ketones are reabsorbed by the proximal tubules of the kidney. When the filtered loads of these substances exceed their T$_m$s, however, they will appear in the urine and produce an osmotic diuresis.**
>
> **Increased ketone production caused a metabolic acidosis, which stimulated her peripheral chemoreceptors. This action produces a characteristic hyperventilation (Kussmaul breathing). In addition, her excess acid is partially buffered by HCO$_3^-$, which is converted to CO$_2$ and expired, reducing the plasma [HCO$_3^-$]. There is an increased loss of Cl$^-$ in the urine. Thus her plasma chloride is also reduced, having been replaced by the anionic components of the ketones.**

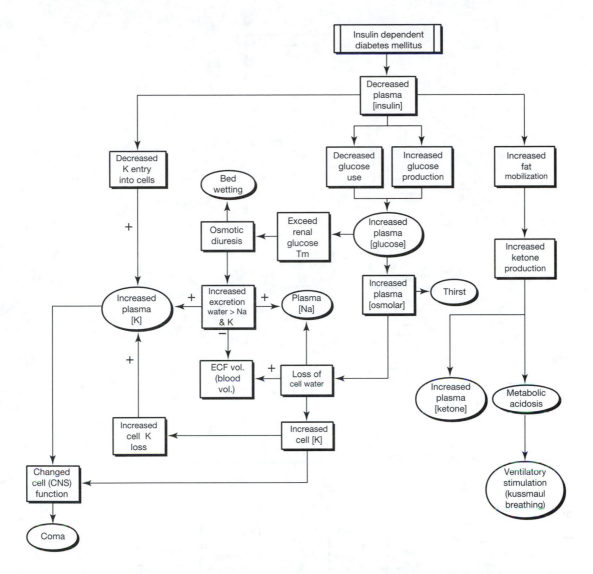

Figure endo 3-2. The source of the signs and symptoms of uncontrolled diabetes mellitus. See the text for a description.

The patient was treated with insulin and intravenous saline. Initially, she showed some improvement, but 2 hours later she again became stuporous. At that time, she had a blood pressure of 80/50 mm Hg, and her blood chemistries were the following.

BLOOD CHEMISTRY	VALUE
glucose	100 mg/dl
ketone bodies	absent
pH	7.38
K+	2.2 mM

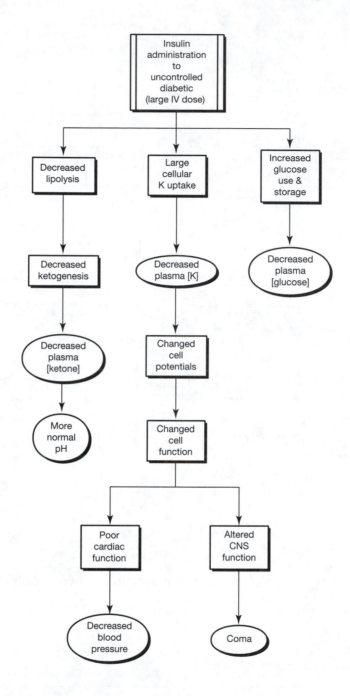

Figure endo 3-3.   The effect of insulin administration to an uncontrolled diabetic. See the text for a description.

5. Explain why she again became comatose.

Insulin directly stimulated cellular uptake of K+, increasing intracellular [K+] while it simultaneously reduced the plasma [K+]. This change in electrolyte distribution hyperpolarized CNS neurons, moving them farther from threshold and making them harder to depolarize, which depressed cellular function (of the heart as well as neurons in the CNS).

The hyperglycemia, the ketonemia, and the acidosis were corrected by the insulin administration (see Figure endo 3-3). Cellular glucose uptake, storage and utilization were all increased. Lipid mobilization was reduced, lipid storage was increased, and ketogenesis was inhibited. Respiratory and renal activity will correct the acidosis in a few days.

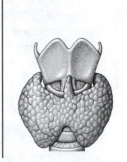

Endocrine 4

HYPOTHYROIDISM

PHYSIOLOGICAL BACKGROUND

The trophic hormones of the anterior pituitary act on peripheral endocrine glands to control their growth and secretion and to sustain their function. The pituitary and the peripheral glands that it controls act in combination with the hypophysiotrophic area of the hypothalamus. The hypothalamus secretes hormones that stimulate and inhibit the individual hormone secretions of the anterior pituitary. Thus together they regulate the plasma concentrations of the peripheral gland hormones (see Figure endo 4-1).

As the figure illustrates, this regulation very much depends on the negative feedback of the peripheral gland hormones on the pituitary and/or the hypothalamus. The negative feedback is reduced when the circulating concentration of the peripheral gland hormones falls, thus allowing the pituitary and hypothalamic secretion rates to rise. Conversely, pituitary and hypothalamic secretion is inhibited when the plasma concentration of the peripheral gland hormones rises. The changes in the secretion rate of the hypothalamic hormones and the pituitary trophic hormones then restores the plasma free peripheral endocrine gland hormone concentration toward or to normal. If the change in the plasma free hormone concentration is persistent, a chronic change in trophic hormone secretion rate will result and will affect the size of the peripheral endocrine gland. If increased trophic hormone secretion rate is sustained, the gland will grow; at a lower secretion level, the gland will atrophy.

CLINICAL BACKGROUND

Excess growth of the thyroid gland, called a goiter, is palpable and/or visible. A common cause of such swelling is a primary decrease in thyroid hormone production due to a dietary iodide deficiency. Any chronic decrease in plasma thyroid hormone concentration, however, could have the same effect. The decrease in thyroid hormone production reduces the feedback inhibition of TSH secretion normally provided by the thyroid hormone. Thus there is an increased TSH output from the pituitary. A chronic increase in TSH secretion causes the thyroid gland to grow excessively, to develop into a goiter.

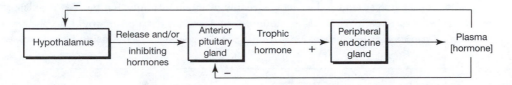

Figure endo 4-1. The endocrine system represented is typical of those with a peripheral gland that is under the combined control of the hypothalamus and the anterior pituitary.

A goiter is also present in some cases of excess thyroid secretion, but the cause for the glandular growth is different in the two cases. Graves disease is a hyperthyroid condition that is often accompanied by goiter. About 80% of patients with Graves' disease have a plasma factor, long-acting thyroid stimulator (LATS), which increases the thyroid secretion rate and stimulates growth of the gland. LATS, a stimulating antibody, is thought to result from an autoimmune reaction against the thyroid gland. The consequent increase in plasma thyroid hormone concentration inhibits pituitary TSH secretion in Graves' disease.

PROBLEM INTRODUCTION

Endocrine glands are components of control systems which function by negative feedback. These systems oppose the actions of disturbances that would tend to disrupt the normal functions that the system regulates. This problem provides an example of such interactions.

Five years ago, Ms. T.H., a 32-year-old secretary, was referred to you by her gynecologist. The referring physician had found an enlarged thyroid during routine examination. Upon questioning, the patient reported none of the symptoms of thyroid disease. She did say, however, that both her sister and her mother took thyroid medication, but for unknown reasons. Physical examination revealed that her thyroid was indeed about twice normal size.

Blood samples were drawn for the measurement of TSH and free T_4.

1. Predict the outcome of these measurements relative to normal (I = greater, D =lower, 0 = normal).

HORMONE	VALUE
TSH	I, D, OR O
free T_4	I, D, OR O

Without more data, you cannot tell why her thyroid gland is enlarged. The enlargement could be the result of excess stimulation by an abnormal agent, as in Graves' disease. Long-acting thyroid stimulator (LATS), an immune agent that stimulates the thyroid to secrete more hormone, is present in most of these patients. The consequent increase in thyroid hormone secretion would raise the plasma thyroid hormone concentration and inhibit TSH secretion by the pituitary gland (see Figure endo 4-2).

The patient could have some problem with her thyroid gland that reduced its hormone output (hypothyroidism). This reduced output would remove some of the feedback inhibition of the anterior pituitary on TSH secretion that is normally caused by the free circulating thyroid hormone. TSH secretion would therefore be elevated and would stimulate thyroid gland growth.

Because the patient exhibits none of the signs and symptoms of thyroid disease it is probable that her plasma free T_4 concentration is still in the normal range. Therefore, regardless of which of the two situations described above is present, the patient is in the early phase of that condition. Her goiter would then be the result of direct stimulation by TSH resulting from a chronic small decrease in thyroid hormone production. Or, her goiter would be the result of excess stimulation by an immune agent having TSH-like properties. The stimulation, however,

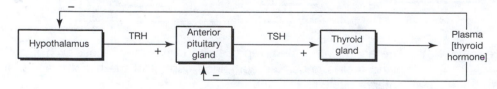

Figure endo 4-2. The system that regulates the plasma free thyroid hormone concentration. It is typical of systems under hypothalamic and anterior pituitary control.

would only increase her thyroid hormone concentration to a high normal value, not enough to produce the clinical signs and symptoms of hyperthyroidism.

The results of the plasma hormone study are shown below.

HORMONE	VALUE
TSH	4.1 mU/L (normal 0.4–4.2)
free T_4	0.9 ng/dl (normal 0.8–2.7)

2. What is the cause of her enlarged thyroid gland?

The patient's TSH concentration is on the high side of normal, and her T_4 is on the low side of normal. Those data support a hypothesis of primary thyroid gland secretion deficiency. If the patient had Graves' disease and LATS production, her T_4 would be high (the effect of LATS) and would be inhibiting her TSH production (TSH would be low).

With primary thyroid disease, the chronically low normal T_4 removes part of the feedback inhibition of the pituitary. TSH secretion is increased enough to partly counteract whatever is causing the reduced T_4 secretion, keeping the plasma concentration in the normal range (although probably lower than it had been before the onset of the disease). The chronically elevated TSH secretion has also stimulated thyroid gland growth.

It is now 5 years later, and Ms. T.H. has come to see you again. You have not seen her in the intervening period. She now complains of a 10-pound weight gain. She also complains of tiredness, sensitivity to cold, and heavier than usual menses. On examination you find that her thyroid is about three times normal size. Her skin is somewhat dry. Her pulse is 56, lower than at her last examination.

A blood sample was drawn for hormone measurements.

3. Predict the outcome, relative to the measurements made five years ago (I = higher than before, D = lower than before, 0 = the same as before) of her plasma hormone concentrations. Explain the basis for your predictions.

HORMONE	VALUE
TSH	I
free T_4	D

Ms. T.H. is now exhibiting symptoms of hypothyroidism: weight gain, slow pulse, dry skin, tiredness, cold sensitivity, menorrhagia. Hence, it is likely that the borderline disease she exhibited 5 years earlier has progressed. Her reduced plasma T_4 would have further "disinhibited" her anterior pituitary, causing a larger increase in TSH output.

The results of her hormone study are given below.

HORMONE	VALUE
TSH	18.6 mU/L (normal 0.4–4.2)
free T_4	0.6 mg/dl (normal 0.8–2.7)

In addition, her antithyroid peroxidase antibodies were determined to be 142.6 IU/ml (normal <2.0).

4. What is the cause of Ms. T.H.'s thyroid problem?

The high antithyroid peroxidase titer indicates that she has an autoimmune thyroid disease, most likely Hashimoto's thyroiditis. This disease destroys thyroid gland cells, reducing the number of secreting thyroid cells and causing hypothyroidism. The thyroid gland cells that are disrupted release peroxidase, which acts as an antigen.

5. Why is she now exhibiting the symptoms of hypothyroidism?

With increasing loss of thyroid gland cells, the glandular T_4 output and the plasma free T_4 concentration have fallen below normal levels.

6. Why has her thyroid size increased still further?

Her thyroid size has increased for exactly the same reason that it grew in the first place. The additional fall in plasma free T_4 removed more of the normal feedback inhibition of the anterior pituitary. Pituitary TSH secretion increased further, causing additional growth of the remaining thyroid tissue.

7. Draw a concept map showing the causal sequence of Ms. T.H.'s medical history.

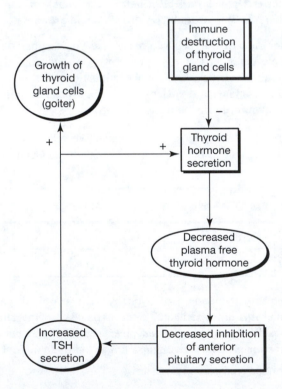

Figure endo 4-3.   A causal diagram of the source of the signs and symptoms that result from immune destruction of thyroid gland cells.

Endocrine 5

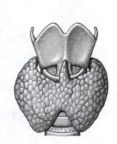

ADRENAL CORTICAL DEFICIENCY

PHYSIOLOGICAL BACKGROUND

The adrenal cortex is organized into three histological layers, each of which secretes a primary hormone having a distinct set of physiological actions. The primary hormone of the outermost layer, the glomerulosa, is aldosterone. Aldosterone is a mineralocorticoid. Its primary target is the kidney where it stimulates Na^+ reabsorption and K^+ secretion. The principle control of aldosterone secretion is by the renin-angiotensin-aldosterone system. ACTH does, however, have a small, transient action. Aldosterone is of primary importance in the regulation of Na^+ and K^+ balance and blood volume. It produces renal salt and water retention in volume depletion. Reduced secretion increases salt and water excretion.

The main hormone of the middle layer, the fasciculata, is cortisol, a glucocorticoid. Its actions oppose almost all the effects of insulin (and are therefore called counterregulatory) and affect the metabolism of glucose, proteins, and fats. The coordinated secretion of insulin and the counterregulatory hormones regulates plasma glucose concentration and ensures the appropriate storage of fuel materials during feeding and the mobilization of stored fuels during food deprivation. Cortisol also has a mineralocorticoid action that is much weaker than that of aldosterone. The secretion of cortisol is controlled by ACTH.

The inner layer of the adrenal cortex, the reticularis, secretes both cortisol and androgens. These secretions are under control of ACTH.

ACTH is synthesized as part of a large molecule, opiomelanocortin. This precursor molecule may be hydrolyzed to produce ACTH, melanocyte stimulating hormone, (MSH), β-lipotropin, and other bioactive compounds.

CLINICAL BACKGROUND

Loss of adrenal cortical function, Addison's disease, results from destruction of the gland by an infectious agent or by autoimmune-induced atrophy. The disease is characterized by weakness, loss of appetite, and the inability to maintain adequate blood glucose concentration during fasting. The patient cannot rapidly excrete water or withstand stress without going into shock. Cortisol deficiency also reduces or removes feedback inhibition of the anterior pituitary, resulting in an increased secretion of ACTH and MSH, both part of the opiomelanocortin precursor. An excess MSH secretion leads to melanin deposition in the skin and mucous membranes.

Decreased aldosterone secretion reduces the kidney's ability to retain Na^+ and excrete K^+. As a result, extracellular fluid volume (including blood volume) falls, leading to orthostatic hypotension, and a decrease in stroke volume, cardiac output, and arterial pressure. Also, if plasma K^+ rises too high, there will be changes in the electrocardiogram; with extreme hyperkalemia (>7 mM), there may be cardiac arrest.

Loss of adrenal androgens is of no consequence in males. However, females show a decrease in the growth of pubic and axillary hair.

PROBLEM INTRODUCTION

A patient often exhibits signs and symptoms that could have resulted from any of a number of causes. To determine the specific cause that is operating in a particular individual requires that the student know the possible causes and know what to measure that will allow these causes to be differentiated. The subject in this problem exhibits an assemblage of symptoms that could arise from a variety of causes.

Ms. A.C. is a 44-year-old lawyer who was referred to a physician by her dentist, who noted hyperpigmentation of her gums. In any case, she had been meaning to see a doctor because of weakness, fatigue, and lightheadedness. She also stated that she had read an article in a magazine that convinced her that she was suffering from hypoglycemia: she develops a rapid heartbeat, shakiness, sweating, and faintness if she does not eat promptly on arising in the morning or if she misses a meal. These symptoms are relieved 15 to 20 minutes after eating. It is January in Chicago, yet the patient, who has not traveled, looks as though she just returned from a vacation in the Caribbean.

1. What are the possible causes of "lightheadedness" or "faintness"?

   These terms are applied to dizziness. There are four principle causes for dizziness. (1) In a subject who experiences a rotational sensation, vertigo, the most common cause is vestibular disease. (2) A sensation of an impending faint most commonly results from some cardiovascular cause that reduces brain perfusion (e.g., orthostatic hypotension, cardiac arrhythmia, carotid sinus hypersensitivity, Valsalva maneuver) or an inadequate supply of energy substrate to the brain, (e.g., hypoglycemia). (3) Vague lightheadedness results from hyperventilation (the hypocapnia causes cerebral vasoconstriction), depression, or anxiety. (4)Disequilibrium upon walking is generally the result of a motor control disorder.

2. What are the possible hormonal and metabolic causes of hypoglycemia?

   Excess production of insulin (insulinoma) or of an insulin-like material (non suppressible insulin-like agent) produce hypoglycemia independently of food intake. Postfeeding insulin secretion overshoot (alimentary insulin excess) is another cause. Glucose regulation depends on normal hepatic function. Hence, inability of the liver to mobilize liver glycogen or produce glucose (gluconeogenesis), as occurs with liver defects (inadequate glycogen storage or lack of hepatic enzymes), causes hypoglycemia. Finally, a deficit of counterregulatory (to insulin) hormones prevents hepatic glucose mobilization or production (panhypopituitarism, decreased growth hormone production, decreased cortisol production).

   Finally, many diabetics try to achieve close glucose control so as to delay the onset of the secondary consequences of the disease. This plan can lead to frequent hypoglycemia from insulin overdose in insulin-dependent diabetics and diabetic drug overdose in non–insulin-dependent diabetics.

3. What could be the cause of Ms. A.C.'s symptoms upon missing a meal?

   The combination of rapid heartbeat, shakiness, and sweating is the classical consequence of hypoglycemia. These symptoms are caused by sympathetic activation induced by a fall in blood sugar.

4. What pathophysiological mechanism(s) might be responsible for the signs and symptoms exhibited by Ms. A.C.? What data in the patient's description lead you to this hypothesis (these hypotheses)?

   1. *Excess secretion of insulin or an insulin-like agent. The symptoms she experiences if she misses a meal are typical of insulin-induced hypoglycemia.*

2. _Deficit in cortisol secretion._ Primary cortisol deficit is associated with increased MSH secretion, accounting for her hyperpigmentation. Low plasma cortisol concentration would also interfere with her ability to mobilize glucose.

3. _Aldosterone deficiency._ The resulting decrease in blood volume could account for her lightheadedness.

5. What would you do to test your hypothesis (hypotheses)?

Measuring her fasting blood glucose would tell us if she were indeed hypoglycemic. That would test hypotheses 1 and 2.

Measuring her serum electrolytes would help determine if her lightheadedness were the result of hypoaldosteronism or perhaps of cortisol deficit. Cortisol does have mineralocorticoid actions.

Measuring her plasma hormone concentrations (insulin, cortisol, aldosterone) would indicate if any were present in smaller than normal concentration.

A standard blood workup was done. The table below shows the significant data.

TEST	RESULT
Na+	132 mM (normal 135–145)
K+	5.2 mM (normal 3.5–5.0)
fasting blood glucose	60 mg/dl (normal 70–120)

6. Do these data support or refute your hypothesis (hypotheses) about what is wrong with Ms. A.C.? What leads you to that conclusion?

The data, fasting hypoglycemia plus appropriate electrolyte changes, support an adrenal cortical problem. Excess insulin production while lowering plasma glucose concentration would tend to increase cellular K+ uptake, thereby reducing plasma [K+].

7. What would you do to test your hypothesis (hypotheses)?

Measure plasma cortisol, aldosterone, and ACTH levels.

The following additional laboratory tests were performed.

TEST	RESULT
morning ACTH	148 pg/ml (normal 8–52)
morning cortisol	6 μg/dl (normal 10–20)

Cortisol was unchanged 1 hour after injecting 0.25 mg of synthetic 1–24 adrenocorticotropin (ACTH) (normal 2 × increase).

8. Why were the hormone measurements made in the morning?

ACTH and cortisol exhibit a diurnal rhythm. The secretion rate and plasma concentrations are highest upon arising and lowest before going to bed. To make the data interpretable, it has become standard to measure the hormone concentrations first thing in the morning.

9. What's wrong with Ms. A.C.?

She has a primary cortisol secretion deficit (low plasma cortisol concentration but high ACTH). She probably is also secreting aldosterone at a lower than normal rate.

10. Draw a concept map that shows the causal relationships between Ms. A.C.'s disorder and her signs, symptoms, and laboratory values.

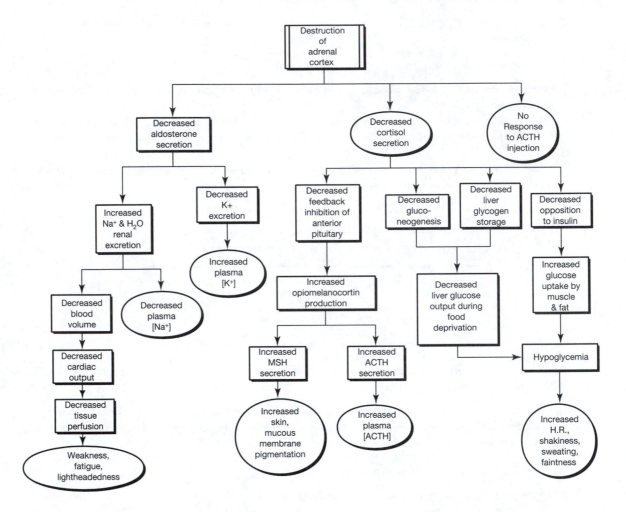

Figure endo 5-1. A concept showing the causal relationships between Ms. A.C.'s hypoadrenalcortical function and her signs, symptoms, and laboratory values. See the answer text for an explanation.

Endocrine 6

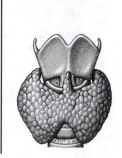

PHEOCHROMOCYTOMA

PHYSIOLOGICAL BACKGROUND

Hormones have widespread actions, influencing the function of all organ systems, including the cardiovascular system. These inputs are part of the extrinsic controls that act on the heart and the blood vessels. Among the important extrinsic influences that affect cardiovascular function are those of the sympathetic nervous system and the secretions of the adrenal medulla. These neurohumoral inputs affect the cardiac output (CO) through their effects on the heart rate and cardiac contractiltity (inotropic state), and they affect the total peripheral resistance (TPR) through their actions on the vascular smooth muscle. Because CO and TPR are the determinants of blood pressure (MAP = CO × TPR), variations in sympathetic activity or adrenal medullary secretion have a powerful impact on blood pressure and its regulation.

Other factors that affect the performance of the heart and/or the activity of vascular smooth muscle will also influence the blood pressure, but by directly or indirectly affecting CO or TPR. For example, changes in blood volume will directly alter the CO.

CLINICAL BACKGROUND

Hypertension has many causes. Some of these exert their effects through known mechanisms. Some if not most produce hypertension by mechanisms that are not known. Pheochromocytoma, an intermittently hypersecreting adrenal medullary tumor, is rare among the causes of hypertension. Because the tumor secretes an excess of catecholamines that act directly on the heart and the blood vessels, however, the hypertension it produces is readily explainable.

PROBLEM INTRODUCTION

When the heart and blood vessels are stimulated, the blood pressure increases. There are many possible ways in which this can be brought about. This problem deals with a situation in which a subject's condition causes her blood pressure to be labile.

A 30-year-old respiratory therapist went to her doctor asking for information about high blood pressure. She had been told 3 years ago, at the time of her pre-employment examination, that she had mild hypertension. She returned several times for follow-up checks. Her blood pressure was normal on three occasions and moderately elevated on a fourth visit. She was informed that she had "labile" hypertension (i.e., her hypertension was only intermittently increased) and was encouraged to have her blood pressure checked every 6 months or so.

1. What pathophysiological—abnormal physiological—mechanisms could produce hypertension on some occasions and not at others just a few weeks later?

If her blood pressure changes markedly over relatively short time periods—that is, over a period of just a few weeks—the cause for the lability must be something that also can change within that time frame. Therefore, she cannot have a permanent problem. The most obvious factor that can change over a short time interval is some neural mechanism like a change in the activity of the autonomic nervous system. Such a change could directly affect the heart and/or the blood vessels. An endocrine mechanism could also change within a time frame of a month; examples are something that affects blood volume and therefore cardiac output and something that affects the concentration of hormone receptors or their sensitivity. A change in kidney function could also affect blood volume.

She had no further problem until 6 months ago. Then she began to experience "spells" during which she became nauseous (sometimes to the point of vomiting), had a headache, and exhibited nervousness, sweating, and a rapid pulse. A fellow therapist took her blood pressure during several of these "spells." It was usually elevated. On one occasion her blood pressure was 200/140 mm Hg.

2. Can any of the pathophysiological mechanisms you previously identified give rise to the additional signs and symptoms that appear during her "spells"? Which ones? Explain how each could lead to these signs and symptoms.

Nausea, with or without vomiting, and headache are commonly associated with very high blood pressures. One cannot use them to eliminate or include possible causes. But nervousness, sweating, and rapid pulse point to excess epinephrine. This agent has powerful CNS stimulating actions (nervousness), stimulates sweating, and acts directly on the heart to increase heart rate. The ANS might also be implicated.

The physician decided that the patient needed further evaluation. Her previous medical history was unremarkable, as were those of her immediate family members.

On examination she was found to be a normally developed white female in no obvious distress. Her BP was 158/95 mm Hg standing and 160/98 mm Hg sitting. Her pulse was 90/min, and her respiratory rate 16/min. Her temperature was 98.6°F.

The rest of her physical examination was unremarkable. SMA6 and SMA12 (screening blood chemistry tests), complete blood count (CBC, of red and white cells) and urine exams were normal, as were her ECG and chest X ray.

3. Does this additional information enable you to eliminate any of the possible pathophysiological mechanisms you had previously identified? What do you think is wrong with this patient?

The only abnormalities found were increased blood pressure, heart rate, and possibly respiration. If she had a kidney or endocrine problem that had caused her blood (volume and) pressure to be elevated, there would probably have been one or more accompanying electrolyte abnormalities. None was found. The autonomic disfunction hypothesis still looks good.

Because of the patient's history of "spells," it was decided to screen her for a pheochromocytoma (a hypersecreting tumor of the adrenal medulla), although only 1% of hypertensives have the disease for this reason.

4. What could you do to test for the presence of pheochromocytoma? What would you expect to see?

Testing for the presence of a pheochromocytoma can be made by measuring the plasma epinephrine concentration or by measuring the urinary excretion of epinephrine metabolites such as metanephrine or vanillylmadelic acid (VMA).

The patient was instructed to collect her urine for 24 hours. The test results are shown below.

AGENT	TEST RESULT	NORMAL
vanillylmandelic acid (VMA)	11.3 mg	<7.0
metanephrine	5.1 mg	<1.5
catecholamines	632.0 mcg	<135
homovanillic acid	4.3 mg	<10

Computerized tomography of her abdomen revealed bilateral masses in the adrenals, and tumors were removed from both adrenals. Histological examination of the specimens revealed findings consistent with pheochromocytoma.

These tests are consistent with the presence of pheochromocytoma. Homovanillic acid, the primary excretory product of dopamine, is not elevated. Dopamine, however, is not secreted by the adrenal medulla in significant quantities.

5. Draw a concept map explaining the source of her signs and symptoms.

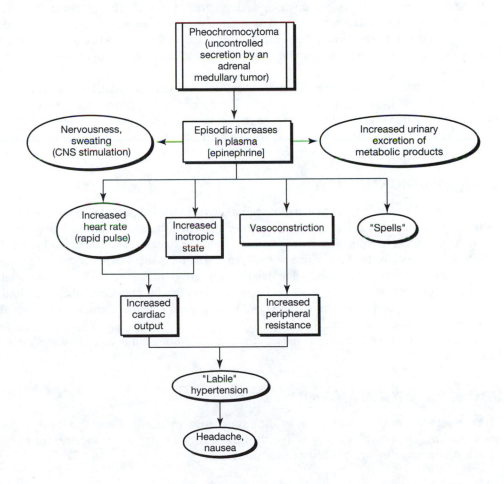

Figure endo 6-1. A causal diagram showing the relationship between the subject's pheochromocytoma and her signs and symptoms.

Gastrointestinal 1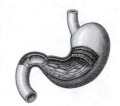

CHRONIC PANCREATITIS

PHYSIOLOGICAL BACKGROUND

The exocrine pancreas produces a juice that is rich in bicarbonate and rich in enzymes essential for digestion. These enzymes include proteolytic enzymes (trypsinogen, which is converted to its active form trypsin in the small intestine; chymotrypsinogen and procarboxypeptidase, both activated in the intestine by trypsin), pancreatic amylase, and pancreatic lipase. The bicarbonate buffers the extremely acidic chyme that enters the small intestine from the stomach and produces a pH more nearly optimum for the function of the pancreatic enzymes. A failure of these exocrine functions will make it impossible for the gastrointestinal tract to digest fully and hence to absorb all the nutrients presented to it.

The endocrine pancreas releases a number of hormones essential for the regulation fuel metabolism (insulin, glucagon, somatostatin). These hormones regulate blood glucose level by altering cell function (uptake of nutrients, conversion to other forms, etc.). A failure of the endocrine pancreas results in the inability to properly use the nutrients absorbed from the gastrointestinal tract (see Endocrine 3).

CLINICAL BACKGROUND

Chronic alcoholism can lead to pancreatitis, an inflammation of the pancreas. This process can lead to tissue destruction and calcification. When the pancreatic lobules are damaged or destroyed, the loss of acinar cells prevents the synthesis and secretion of proteases, amylases, and lipases that are essential for digestion. In the absence of these enzymes, the patient experiences a malabsorption syndrome and undigested or partially digested food is excreted in the feces. The loss of ductal tissue reduces the production of bicarbonate and impacts digestion by preventing the neutralization of stomach acid that reaches the intestine. Thus the pancreatic enzymes, which require an alkaline environment, do not work optimally.

Loss of the islets of Langerhans reduces the availability of insulin, and a diabetes mellitus can be the end result.

PROBLEM INTRODUCTION

The pancreas plays a key role in digestion and in controlling the body's fuel metabolism. Therefore, changes in either the exocrine or the endocrine functions of the pancreas seriously impact homeostasis.

Mr. G.I. is an alcoholic with a long history of chronic pancreatitis. During his most recent hospitalization, X rays taken showed extensive calcified fibrous tissue throughout the pancreas.

1. The *exocrine* pancreas is made up of two structures, each of which contributes to the production of pancreatic juice. List the two structural components and their secretory products in the table below.

STRUCTURES	SECRETORY PRODUCT(S)
ACINAR CELLS	PANCREATIC ENZYMES
DUCTAL CELLS	BICARBONATE RICH AQUEOUS SOLUTION

2. Predict the consequences (increase, decrease, no change) of the loss of pancreatic acinar cells on the digestion and absorption of the following.

	DIGESTION	ABSORPTION
carbohydrates	DECREASE	DECREASE
proteins	DECREASE	DECREASE
fats	DECREASE	DECREASE

3. Explain each prediction.

Loss of pancreatic amylase limits the digestion of complex carbohydrates and the availability of their digestion products, disaccharides, for further digestion and then absorption. The undigested carbohydrates increase the osmotic pressure of the luminal contents, reducing water absorption from the gut and leading to diarrhea.

Loss of pancreatic proteases prevents the digestion of proteins and the absorption of their digestion products, amino acids. The result will be a negative nitrogen balance.

Loss of lipases prevents the digestion of fats and the absorption of their digestion products, fatty acids. The undigested fat remains in the lumen, producing a fat-rich stool, a condition called steatorrhea.

4. Predict the consequences (increase, decrease, no change) of the loss of pancreatic ductal tissue on the digestion and absorption of the following.

	DIGESTION	ABSORPTION
carbohydrates	DECREASE	DECREASE
proteins	DECREASE	DECREASE
fats	DECREASE	DECREASE

5. Explain each prediction.

Loss of ductal tissue reduces the secretion of bicarbonate. Thus, when the stomach contents, which have a very low pH, enter the duodenum, the duodenal contents remain acidic. The pancreatic enzymes have optimal pHs in the alkaline range, and in an acidic environment they do not work as well as they normally would.

The diagnosis of chronic pancreatitis is often made by delivering physiological stimuli to the pancreas by intravenous administration of some agent and then measuring the response using a catheter placed in the duodenum to collect samples of pancreatic juice.

6. What agent would you use to stimulate enzyme secretion? Explain.

Administer CCK intravenously and measure the rate of enzyme secretion. Note, however, that fluid and bicarbonate secretion rates will also be increased.

When presented with a patient showing clinical signs of pancreatitis, another test—measurements of serum amylase and lipase—is also used. The inflammatory process at work causes "autodigestion" of pancreatic tissue, which liberates the enzymes contained in the acinar cells into the tissues from where they pass into the blood.

7. What agent would you use to stimulate fluid secretion? Explain.

Secretin administered intravenously would increase the volume and bicarbonate content of pancreatic secretions. In the presence of damage to the pancreatic ducts, this secretion would be decreased.

The pancreas also exhibits important *endocrine* functions. Pancreatitis can also affect these functions.

8. How might Mr. G.I.'s general metabolism be altered by a loss of pancreatic endocrine function?

The inflammatory process can also destroy the islets of Langerhans. Loss of β-cells and their secretion, insulin, can significantly impact glucose, fat, and protein metabolism, producing symptoms of decreased glucose tolerance or even frank diabetes mellitus. (See Endocrine 3 for a discussion of diabetes mellitus.)

Gastrointestinal 2

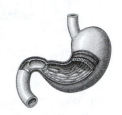

CONTROL OF GASTRIC EMPTYING ("DUMPING" SYNDROME)

PHYSIOLOGICAL BACKGROUND

Gastric motility and emptying are controlled so that the processes carried out in the stomach have time to be completed and the preparation for processing in the small intestine can begin. The emptying of the stomach is controlled by neural and hormonal signals that modulate gastric smooth muscle and pyloric sphincter contraction. When chyme enters the duodenum, chemoreceptors respond to acid, the products of fat digestion, and hypertonicity and send signals to the intramural intrinsic nerve plexuses that in turn send signals to the CNS and directly to gastric smooth muscle. The resulting reflex slows gastric emptying. In addition, acid in the duodenum releases secretin, fats release GIP and CCK, and hypertonicity releases a presently unknown hormone, which all slow gastric emptying. The result is that the duodenum receives chyme at a rate that allows it to carry out its digestive function.

When the regulation of emptying is disturbed and the stomach empties much more rapidly, the duodenum receives chyme faster than it can process it. The results are duodenal distention and hypertonicity of duodenal contents. As a consequence, fluid shifts from the vascular compartment to the lumen of the gut. The symptoms associated with the "dumping" syndrome arise from this GI pathophysiology.

CLINICAL BACKGROUND

Peptic ulcers are common. Whatever the underlying cause of ulcers (and there is recent evidence that a bacterial infection is involved), the mechanisms involving hypersecretion of acid and the breakdown of the normally protective mucosal barrier are well understood. There are a number of approaches to treating this condition, but when perforation of a gastric ulcer has occurred, part of the stomach is removed, the vagal innervation of the stomach may be sectioned, and the pyloric valve between the stomach and the duodenum may be removed. A not infrequent consequence of such surgery is the occurrence of postprandial dumping. The early symptoms of this condition include GI distress and vasomotor disturbances (flushing, dizziness, tachycardia). Later symptoms appear to arise from a hypoglycemic state.

Treatment for this condition is usually dietary. The patient should eat frequent, small meals (thus reducing distention of the stomach and slowing emptying) that are high in fat and protein and low in carbohydrates (reducing the stimuli that promote emptying).

PROBLEM INTRODUCTION

The sequence of events that make up normal digestion requires the controlled movement of the food being processed through the gastrointestinal tract. If the timing of these events is disrupted the consequences can be grave.

Mrs. Bilious is a 50-year-old woman with a history of juvenile onset arthritis. Management of her condition has included the use of large doses of aspirin and nonsteroidal anti-inflammatory drugs (NSAIDs). Recently she developed a gastric perforation that required removal of most of the body and all the antrum of the stomach (subtotal gastrectomy) and surgical intervention to facilitate gastric emptying (pyloroplasty, or surgical widening of the pylorus, and attachment of the remaining stomach directly to the jejunum).

1. Why did Mrs. Bilious develop a gastric ulcer that eventually perforated?

   Aspirin and NSAIDs inhibit prostaglandin synthesis and thereby decrease the secretion of mucus and bicarbonate in the gastric mucosal barrier of the stomach. As a result of prolonged use and/or high doses of these drugs, the protection against the destructive effects of gastric secretions afforded the mucosa of the stomach by the normal layer of bicarbonate-rich mucus is lost. Once the barrier is compromised, the extremely low pH of the stomach damages the cells, releasing histamine. Histamine stimulates further secretion of acid. The end result of this process is an ulcer. If the damage is extreme, a hole or perforation can appear in the wall of the stomach. Eroded blood vessels can hemorrhage, and the gastric contents can enter the peritoneal cavity.

2. What effect (increase, decrease, no change) will the surgical procedures she has undergone have on the rate of gastric emptying? Explain.

   Bypassing the normal function of the pyloric valve (whether by pyloroplasty or gastrojejunostomy) results in the loss of the feedback (neural and hormonal) that normally slows gastric emptying. As a result, the stomach delivers chyme to the intestine at a rate greater than the duodenum can process it. This process leads to distention of the intestine (which can induce vomiting), incomplete digestion, and hence incomplete absorption, and osmotic fluid shifts from the vascular compartment into the lumen of the intestine because of the high concentration of osmotically active solutes in the intestinal contents.

Gastric emptying will occur much more rapidly than normal, producing what is known as a "dumping" syndrome. Early symptoms of this condition include GI distress and vasomotor disturbances.

3. What will be the consequences of "dumping" on the functions of the duodenum? Explain.

   The pH of the intestine will be decreased (it will become more acidic). Pancreatic juice, which normally neutralizes the stomach contents as they enter the duodenum, will not be able to maintain a pH appropriate for the action of the pancreatic enzymes (the buffering by HCO_3^- will be insufficient to neutralize the amount of acid present), and digestion will be incomplete.
   There will also be an abnormally large osmotic load in the intestine, which will cause large shifts of water from the vascular compartment into the lumen of the intestine. Thus the large amount of relatively undigested food will be propelled rapidly through the GI tract along with this water, resulting in diarrhea.

4. What possible cardiovascular symptoms might be a part of the "dumping" syndrome? Explain.

The osmotic shift of water from the vascular compartment into the lumen of the GI tract will result in decreased blood volume. With central venous pressure reduced, there will be decreased filling of the ventricles and hence cardiac output will decrease. The resulting fall in mean arterial pressure will cause a baroreceptor reflex to occur, producing tachycardia and peripheral vasoconstriction. If the shift in fluid is large enough, hypovolemic shock may result.

5. How might the patient's problem be managed most simply?

A simple prescription for this patient would be to eat small, frequent meals. In essence, the patient is then behaviorally metering the emptying of the stomach into the intestine at a rate that the intestine is able to accommodate.

Gastrointestinal 3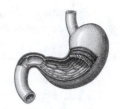

MALABSORPTION SYNDROME

PHYSIOLOGICAL BACKGROUND

Almost all digestion (the breakdown of the large molecules making up the ingested food into smaller molecules) and absorption of the products of digestion take place in the intestine. The structure of the intestine, from the enterocyte to the mucosa as a whole, supports both of these functions.

The digestion of carbohydrates (sugars, starches), proteins, and fats is accomplished by enzymes secreted by the pancreas (proteolytic enzymes, pancreatic amylase, lipases) and carried to the small intestine in the pancreatic duct. In addition, enzymes located in the brush border of the intestinal villi (disaccharidases and aminopeptidases) contribute to digestion.

The highly folded structure of the intestinal mucosa presents a very large surface area of epithelial cells across which absorption can occur. Some of the monomeric nutrients are absorbed by passive diffusion and some by active transport processes. Damage to the intestinal mucosa, by interfering with digestion (by loss of the brush border enzymes), and absorption (reduced surface area), can give rise to malabsorption syndromes.

CLINICAL BACKGROUND

Adult celiac disease (nontropical sprue) is a common cause of malabsorption syndrome. This condition is caused by an autoimmune reaction to dietary gluten (or one of its components), a high-molecular-weight protein found in all grains except corn and rice. The toxic response damages the villus cells of the intestinal epithelium, and characteristic histological changes occur. The net result is a marked decrease in the epithelial surface area across which absorption of the products of digestion can occur. Secondarily, damage to the enterocytes can decrease the brush border enzymes (peptidases and amylases) that normally contribute to the final phases of digestion of proteins and carbohydrates. Absorption of fat is impaired by the decreased absorptive surface area.

Because the etiology of this condition is still unknown, the only treatment is a diet that is totally free of gluten.

PROBLEM INTRODUCTION

The structure of the intestinal mucosa, both at the histological level and at the level of gross structure, is important for intestinal function. Abnormalities of structure will therefore result in altered intestinal function and disturbances to whole body homeostasis.

The mucosa of the small intestine is a heavily folded structure. It consists of villi projecting into the intestinal lumen lined with the epithelial cells. The epithelia have microvilli on the luminal surface. Figure gi 3-1 illustrates the normal structure of the intestinal mucosa.

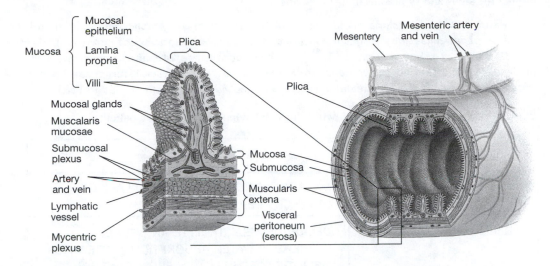

Figure gi 3-1. The normal structure of the intestinal mucosa (from Martini, *Fundamentals of Anatomy and Physiology*, 4th ed., Prentice Hall, 1998) p. 864.

1. What two functions does intestinal mucosa serve in the overall process by which ingested food is made available to the cells of the body?

   The highly folded intestinal mucosa has a very large surface area, a feature that results in a very large absorptive surface area. The microvilli further increase the area across which transport processes (passive or active) can move nutrients to be absorbed.

   In addition, the microvilli membranes contain intrinsic digestive enzymes, particularly peptidases and amylases, which play an important role in the final stage of digestion of proteins and carbohydrates, respectively. Note, however, that the brush border enzymes do not include lipases.

Pathology of the intestinal mucosa alters intestinal function; both digestion (the breaking down of complex molecules into smaller molecules) and absorption into the blood will be abnormal. Consider an individual whose intestinal villi have been altered by a tissue reaction to something ingested so that villi are flattened and the microvilli are shortened and blunted.

2. How will this pathology affect intestinal function? Explain.

   The structural damage described will seriously reduce the area available for absorption and thus will reduce the uptake of any products of digestion that are present in the intestinal lumen. In addition, the loss of brush border enzymes will reduce the digestion (and eventual absorption) of proteins and carbohydrates.

3. Will all nutrients be equally affected? If not, what will be affected most? Why? Explain.

   Absorption of fats is more affected than absorption of proteins and carbohydrates, presumably as a result of the decrease mucosal area for absorption (brush border enzymes do not include lipases). Vitamins, particularly fat-soluble ones, will be affected more than water-soluble ones. Absorption of all electrolytes will be affected.

Gluten-sensitive enteropathy (sprue) is a condition in which the intestinal mucosa is damaged by a response to one of the molecular components of gluten. Sprue is one example of a malabsorption disease.

4. Predict the signs and symptoms of a sprue patient.

Signs and symptoms include fat in the stool (steatorrhea, due to the failure to absorb the products of fat digestion), weight loss (due to the reduced absorption of nutrients), diarrhea (due to damage to the small bowel, leaving it in a net secretory state, and the osmotic retention of water in the intestine due to nonabsorbed nutrients), electrolyte disturbances (lost in the watery stool), and signs of vitamin deficiencies (ingested vitamins are not absorbed).

5. Explain the cause of each of the following signs and symptoms of a sprue patient.

SIGNS/SYMPTOMS	CAUSE
high fat content in stool	MALABSORPTION OF FAT DUE TO LACK OF SURFACE AREA
diarrhea	DECREASED ELECTROLYTE AND HENCE DECREASED WATER ABSORPTION; INCREASED OSMOTIC LOAD IN LUMEN FROM GENERAL MALABSORPTION
weight loss	INABILITY TO ABSORB NUTRIENTS PRESENT IN DIET
weakness and fatiguability	ELECTROLYTE IMBALANCE AND EFFECT ON MUSCLES
anemia	DECREASED ABSORPTION OF IRON
tetany and parasthesias	HYPOCALCEMIA (DUE TO FAILURE TO ABSORB Ca AND VITAMIN D) AND CONSEQUENT HYPEREXCITABILITY OF SENSORY AND MOTOR NERVES

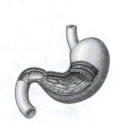

Gastrointestinal 4

ACID PRODUCTION IN THE STOMACH AND ULCERS

PHYSIOLOGICAL BACKGROUND

Digestion of protein begins in the stomach. Pepsinogen, an inactive form of the enzyme pepsin, is secreted by the chief cells. HCl secreted into the gastric lumen by parietal cells cleaves a small polypeptide from pepsinogen, giving rise to the active form of the enzyme. The very acidic conditions present in the stomach (as low as pH 2.0) also are required for optimal pepsin activity.

The same intensely acidic conditions in the stomach pose a serious threat to the integrity of the tissue there. Normally, the gastric mucosa is protected from the low pH by a bicarbonate-rich layer of mucus that neutralizes the excess H^+ ions that diffuse to the surface.

When the stomach empties, the intensely acidic chyme that reaches the duodenum stimulates pancreatic secretion. The pancreatic juice contains a high concentration of bicarbonate (alkaline pH) that neutralizes the acid of the chyme in the duodenum.

CLINICAL BACKGROUND

An ulcer is an erosion (a breakdown) in the wall of either the stomach (a gastric ulcer) or the small intestine (a duodenal ulcer). Damage to the mucosa gives rise to pain, and if the damage is extensive enough, blood vessels may be injured and profuse bleeding can result. If the damage to the wall is still more severe, the ulcer can perforate, allowing the stomach or duodenal contents to enter the peritoneal cavity.

Although there is no question that damage to the mucosa results from the action of acid on the tissue, the question of the mechanism by which the normal protection becomes ineffective is not completely settled.

Ulcers are relatively common, with considerable variability according to age and gender, Duodenal ulcers are much more common than gastric ulcers.

PROBLEM INTRODUCTION

The enzymes that work in the stomach require a very acidic pH for their optimum functioning. Such a low pH, however, is potentially very damaging to tissue, so the stomach (and the small intestine) must be protected from the dangerously low pH that is normally present in the stomach following the ingestion of a meal. The nature of the protective mechanisms involved and the consequences of disrupting this protection are considered here.

While digesting a meal, the pH of the stomach can get as low as 2.0.

1. Describe the mechanism by which the stomach secretes acid and achieves a pH of 2.0.

   To achieve a pH of 2.0 in the stomach requires an active transport process for H^+ on the lumenal membrane of the parietal cells. The H^+ that is pumped into the lumen is derived from CO_2 that hydrates (a reaction promoted by the presence of carbonic anhydrase) and then dissociates into H^+ and HCO_3^-. The bicarbonate that is produced is exchanged for Cl^- across the basal membrane (giving rise to the "alkaline tide" that leaves the stomach in the venous effluent following the ingestion of a meal). Chloride enters the lumen passively.

2. What stimuli control the secretion of acid in the stomach?

   Gastric acid is secreted by the parietal cells located in the oxyntic region of the stomach. The principle stimuli for secretion of gastric acid are acetylcholine (ACh) and gastrin. Histamine, which is released from damaged tissue (i.e., ulcer) is also a potent stimulator of acid secretion. Inhibition of acid secretion is brought about by low pH, secretin, GIP, and CCK. These interactions are described in Figure gi 4-1 p. 365.

3. How is the stomach itself protected from the very acidic conditions that are present?

   The stomach is protected by a bicarbonate-rich mucous lining that prevents the high concentration of H^+ ions from affecting the mucosal cells. In addition, the mucosal cells turn over very rapidly, so no cell is exposed to the low pH for very long. It is also important that the luminal cell membrane is impermeable to H^+, thus preventing the H^+ from entering the cells and causing damage.

4. How is the duodenum protected from the acidity of the chyme that enters it as the stomach empties?

   The duodenum is protected by the buffering capacity of the bicarbonate-rich pancreatic secretion that enters the intestine. This secretion is stimulated by secretin that is released from cells in the duodenal mucosa in response to the acid reaching the duodenum in the chyme.

A gastric ulcer is a pathology in which conditions in the stomach result in cell damage or death and hence the erosion of the wall. The blood vessels of the stomach wall will also be damaged, and bleeding can thus result.

5. By what mechanism can an ulcer result from:
   a. prolonged and/or high doses of aspirin?
   b. hypersecretion of gastrin?
   c. stress?

   a. Aspirin damages the mucosal barrier, at least in part by inhibiting the synthesis of prostaglandins that normally stimulate mucus and HCO_3^- production.

   b. Gastrin directly stimulates the secretion of acid by the parietal cells.

   c. Stress results in neural stimulation of the stomach that causes increased secretion of acid. Activity in the vagus nerve directly stimulates release of gastrin from parietal cell. Gastrin directly stimulates release of acid from the parietal cells. In addition, the vagus directly innervates the parietal cells, and the ACh released there is a powerful stimulus for acid secretion.

6. Why is it fair to say that an active ulcer is an example of a positive feedback situation?

   Damage to the wall results in release of histamine, which in turn stimulates the further release of acid by the parietal cells. This is an example of positive feedback.

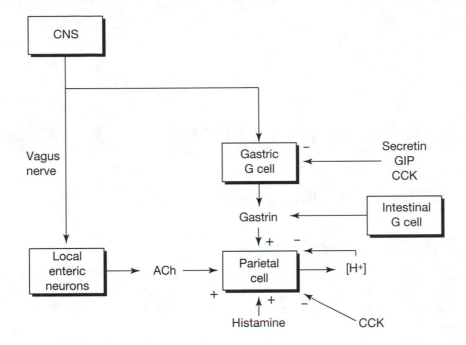

Figure gi 4-1. A concept map representing the control of acid secretion in the stomach.

Patients with Zollinger-Ellison syndrome (which is the result of a tumor of gastrin-secreting cells) exhibit (1) elevated serum gastrin concentrations, (2) increased basal secretion of stomach acid, and (3) peptic ulcer disease and/or diarrhea. Some patients also exhibit steatorrhea (fat in their stool) and evidence of malabsorption.

7. Explain the occurrence of steatorrhea and malabsorption in a Zollinger-Ellison patient.

   Steatorrhea is due to the effect of high acidity on pancreatic lipase. The large acid load presented to the small intestine cannot be neutralized by the alkaline pancreatic secretions, and thus the pH of the small intestine is decreased. The lipases do not work as effectively in an acid environment as they normally would. Undigested fat is present in the lumen and appears in the stool. At the same time, the normal products of fat digestion, monoglycerides and free fatty acids, are reduced; hence, their absorption is reduced.

Gastrointestinal 5

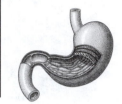

DIGESTION OF A MIXED MEAL

PHYSIOLOGICAL BACKGROUND

The processing of a meal begins in the mouth and proceeds through the stomach, the small intestine, and then the large intestine. It involves the digestion of large complex molecules (starches, proteins, fats) into smaller molecules that can be absorbed from the lumen of the gastrointestinal tract into the extracellular fluid (ECF) and then into the blood. Digestion requires the secretion of appropriate enzymes and the creation of conditions in which these enzymes can function optimally. The pH is perhaps the critical determinant of enzyme activity, and considerable activity occurs in the GI system to ensure that the proper pH is present to allow the digestive enzymes to function optimally.

   Absorption involves passive, active, and secondary active transport processes to move different constituents of the meal into the ECF. In addition, water, electrolytes (most of which must be maintained at more or less constant levels in the body), and vitamins must also be absorbed.

PROBLEM INTRODUCTION

Processing a mixed meal involves a complex sequence of events occurring over a relatively long time at a variety of sites along the gastrointestinal tract. The nature of each step and the processes that control and coordinate each one must be understood.

You go into your favorite Italian restaurant and order a meal of (A) spaghetti and (B) meatballs in (C) marinara sauce with a (D) nondiet soft drink.

   1. Describe the predominant biochemical constituents of each of the components of this meal.

MEAL	CHEMICAL CONSTITUENTS
spaghetti	CARBOHYDRATES
meatballs	PROTEINS, FATS, WATER, ELECTROLYTES
marinara sauce	CARBOHYDRATES, VITAMINS, WATER, ELECTROLYTES
nondiet soft drink	CARBOHYDRATES, WATER

2. Describe the contents of your stomach 1 hour after eating this meal. What, if anything, has happened to the chemical constituents of the meal? What, if anything, has been digested, and what, if anything, has been absorbed?

CHEMICAL CONSTITUENTS	WHAT HAPPENS TO . . . IN THE STOMACH
carbohydrates	SALIVARY AMYLASE CONTINUES DIGESTION
proteins	PEPSIN BEGINS DIGESTION (pH IS LOW; OPTIMUM FOR PEPSIN)
fats	NO DIGESTION OR ABSORPTION
vitamins	NO ABSORPTION
electrolytes	NO ABSORPTION
water	NO ABSORPTION

3. Digestion and absorption of carbohydrates, proteins, and fats are largely completed in the small intestine. Describe the process of digestion and absorption for each of these nutrients in the small intestine.

	DIGESTIVE ENZYMES	WHERE THEY COME FROM	PRODUCTS OF DIGESTION	MECHANISM OF ABSORPTION
carbohydrates	amylase lactase sucrase maltase	pancreas brush border	oligosaccharides disaccharides glucose galactose fructose	into enterocyte by secondary active transport; out to ECF by passive diffusion
proteins	trypsin chymotrypsin peptidases	pancreas pancreas brush border	peptides oligopeptides & amino acids	into enterocyte by secondary active transport; out to ECF by passive diffusion
fats	lipases acting on micelles formed through action of bile salts	pancreas	monoglycerides free fatty acids	into enterocyte by passive diffusion from micelle; in cell chylomicrons are formed and leave the enterocyte by exocytosis and enter the lacteals by diffusion

4. Where and by what mechanism(s) are vitamins, water, and electrolytes absorbed?

	Site of absorption	Mechanism
vitamins	jejunum and ileum	Water soluble (C, B_1, B_{12}) vitamins absorbed into enterocyte by secondary active transport; B_{12} requires intrinsic factor. Fat-soluble vitamins (A, D, E, K) diffuse passively from micelle into cell and leave cell in chylomicrons.
water	jejunum and ileum	Passively down osmotic gradient created by absorption of electrolytes.
electrolytes	jejunum	Na enters enterocyte passively and is pumped out by the Na/K-pump; Cl moves paracellularly down the electrical potential gradient established by Na absorption; K is absorbed passively as water is absorbed; Ca is actively absorbed (stimulated by active forms of vitamin D_3); and Fe is actively transported into cell and leaves passively (transported in blood bound to transferrin). Fe is also stored in cell as ferritin.

Body Fluid Compartments 1

COMPARTMENT SIZE MEASUREMENT

PHYSIOLOGICAL BACKGROUND

The primary constituent of our bodies is water. In a male of average build, about 60% of the body weight is water (called total body water, or TBW). In a female of average build, the figure is about 50%. The difference is largely due to the female's larger body-fat content, which contains very little water (10%).

Total body water is divided into two (conceptual) compartments: intracellular and extracellular. In an individual of average build, about 60% of the TBW is intracellular fluid (ICF) and about 40% is extracellular fluid (ECF).

The technique employed in this problem to determine the body fluid volumes is called indicator dilution and is based on the following concept. One administers to an individual a known amount of some material (Q), called an indicator, that does not affect body fluid content or distribution. One can determine the volume into which the indicator distributes by measuring its plasma concentration (C) and using the relationship

$$\text{concentration } (C) = \frac{\text{quantity } (Q)}{\text{volume } (V)}.$$

Rearranging, one can solve for the volume: $V = Q/C$.

Thus an indicator that is confined to the ECF can be used to measure the size of that compartment, and an indicator that enters the TBW can be used to measure its volume. There are no indicators for measuring the ICF volume. That volume is determined as the difference between the TBW and the ECF volumes (ICF = TBW − ECF).

PROBLEM INTRODUCTION

Excessive gains or losses of water and/or electrolytes cause changes in the body fluid compartments. Such disturbances affect the volume, composition, and concentration of body fluids, and this can have profound effects on many bodily functions.

The distribution of the body fluids of a man weighing 75 kg was being evaluated because of a suspected problem. He was injected with 100 curies (Ci), a measure of radioactivity, of 35 S-sulfate (radiosulfate) intravenously. Twenty minutes after the injection, a blood sample was drawn. It contained 0.0057 Ci of radiosulfate per milliliter of plasma water. In the 20-minute equilibration period, 4 Ci of the injected sulfate was lost in the urine. Assume that there were no other losses and that the tracer had distributed uniformly in the compartment that it entered.

1. Calculate the volume into which the sulfate distributed.

The volume (V) determined with sulfate is often called the sulfate space. Because cell membranes are impermeable to sulfate, it distributes within the extracellular fluid (ECF).
    The sulfate space can be calculated as

$$V \text{ of space} = \frac{Q \text{ remaining in space}}{C \text{ in space}}$$

where Q remaining = Q injected − Q excreted

$$V = \frac{(100 - 4)Ci}{0.0057 \text{ Ci/ml}}$$

$$= \frac{96}{0.0057} \text{ ml}$$

$$= 16,842 \text{ ml}$$

$$\underline{V = 16.8 \text{ L} = ECF \text{ volume}}$$

The same subject was simultaneously injected with 99.8 ml of D_2O (heavy water). Two hours after the injection, a blood sample was drawn. It contained 0.232 g of D_2O per 100 ml of plasma water. In the 2-hour equilibration period, 0.4 g of the injected D_2O was lost in the urine. Assume that there were no other losses and that the D_2O had distributed uniformly in the compartment that it entered.

2. Calculate the volume into which the D_2O distributed.

D_2O, heavy water, enters all the space occupied by ordinary water, the TBW. The same indicator dilution principle is used as in the previous problem:

$$V = \frac{Q \text{ remaining}}{C}$$

$$= \frac{(99.8 - 0.4) \text{ g}}{0.232 \text{ g/100 ml}}$$

$$= \frac{99.4}{0.232} \times 100 \text{ ml}$$

$$= 42,844 \text{ ml}$$

$$\underline{V = 42.8 \text{ L} = TBW \text{ volume}}$$

3. What are the subject's:

COMPARTMENT	VOLUME (L)
total body water	42.8
intracellular fluid	26.0
extracellular fluid	16.8

One cannot directly determine the volume of the ICF using the indicator dilution technique. Instead, the ICF is determined by subtracting the ECF volume from the TBW:

$$ICF = TBW - ECF$$

$$= 42.8L - 16.8L$$

$$\underline{ICF = 26.0L}$$

4. Calculate the fluid volumes for a normal 75-kg man of average build, and enter these values (NORMAL) and the values that were determined for the subject (SUBJECT) in the table below. Calculate and enter the difference between the two sets of values (DIFF).

A man of average build has a TBW volume that is about 60% of his body weight:

$$TBW = 0.60 \times 75 \text{ kg} = 45 \text{ kg} (= 45 \text{ L}: 1 \text{ kg of water} = 1 \text{ L})$$

His ECF is 40% of that:

$$0.40 \times 45.0L = 18.0 \text{ L}$$

The rest is intracellular:

$$ICF = 45L - 18L = 27 \text{ L}$$

COMPARTMENT	SUBJECT (S)	NORMAL (N)	DIFF. (S-N)
total body water	42.8	45.0	2.2L
intracellular fluid	26.0	27.0	1.0 L
extracellular fluid	16.8	18.0	1.2L

5. What could have happened to the subject to produce these values? Explain your answer.

All the subject's body fluid volumes are decreased. Hence, the subject must have lost fluid; he is dehydrated. The question then becomes, What fluid can he have lost? He cannot have lost an extracellular fluid like blood or GI secretions. These fluids are iso-osmotic to body fluids. Their loss would not modify the ECF osmolarity; hence, their loss would not cause an osmotic gradient between the ECF and the ICF.

The subject lost fluid from his ICF, however, and the only way that could have occurred is if an osmotic gradient existed between his ECF and his ICF. The subject's ECF osmolarity would have to have increased to cause water to leave the ICF. Therefore, the subject must have lost a hypo-osmotic fluid, causing the osmotic concentration of his ECF to increase and its volume to decrease. The osmotic gradient would then have caused water to move from ICF to ECF (see Figure co 1-1a). The final outcome would then be a decrease in both ECF and ICF volumes and an increase in both of their osmolarities (see Figure co 1-1b).

There are three places from which one can lose hypo-osmotic fluids. First, insensible perspiration is a pure evaporative water loss by diffusion from body surfaces (not sweat, which is actively secreted by the cutaneous sweat glands). Second, excessive loss of sweat will also concentrate the ECF because sweat is hypo-osmotic. This process, of course, would be caused by exposure to a hot environment and/or an increase in heat production (exercise). Finally, loss of a large volume of dilute urine would also concentrate the ECF. Such a loss could be caused by insufficient ADH secretion or by a loss of kidney responsiveness to ADH.

Without more information, we cannot decide among the three possibilities. It is clear, however, that one or more of them would be responsible for the subject's abnormal body fluid distribution.

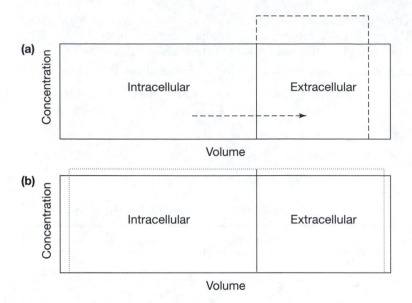

Figure co 1-1. Two Darrow-Yanet diagrams, plots of volume versus concentration. The solid lines show the intracellular (ICF) and extracellular (ECF) fluid volumes and concentration before the loss of a hypo-osmotic body fluid. In (a) fluid loss has reduced the ECF volume and concentrated it (dashed outline), creating an osmotic gradient. Water would then diffuse from ICF to ECF (arrow). Part (b) shows the final steady-state (dotted outline). Water movement has equalized the ECF and ICF concentrations; it has reduced the ICF volume and increased the ECF volume but not to the predehydration volume. The volumes of both compartments are now reduced.

Body Fluid Compartments 2

EFFECTS OF WATER INTAKE

PHYSIOLOGICAL BACKGROUND

With very few exceptions, water freely permeates cell membranes. Therefore, the intracellular fluid (ICF) and the extracellular fluid (ECF) have the same osmolar concentration (C) in the steady-state. Further, since V = Q/C, the volume (V) of each compartment is determined by the quantity of solute (Q) that is present in it.

In individuals of average build, the volume of the total body water (TBW) and its division into ECF and ICF is reasonably constant. Average figures for these are, approximately,

> **TBW = 60% of body weight in males, 50% in females**
>
> **ECF = 40% of TBW**
>
> **ICF = 60% of TBW**

PROBLEM INTRODUCTION

Body fluid volumes are affected by any imbalance between input and output. The volume and composition of the excess intake or loss determine the effect on the size and concentration of the individual fluid volumes. This problem deals with the effects of drinking tap water.

A normal 60-kg man of average build quickly drank 2 liters of tap water. Before this, his osmolar concentration and fluid compartment sizes were normal. Calculate the size and osmolarity of his body fluid compartments before he drank. Also calculate the size and osmolarity of his compartments after he drank and absorbed the water but before he excreted any of it. Enter the results of your calculations in the table on the following page.

In a 60-kg male of average build,

> TBW = 0.6 × 60 kg = 36 kg = 36 L
>
> ECF = 0.4 × 36 L = 14.4 L
>
> ICF = 0.6 × 36 L = 21.6 L

These figures have been entered into the table below.

The normal osmolar concentration is assumed to be 300 mOsm/L. It too has been entered in the table. The osmole content of each compartment is then calculated as: Q (osmoles) = V (L) × C (mOsm/L).

TIME	TOTAL BODY WATER			EXTRACELLULAR FLUID			INTRACELLULAR FLUID		
	VOL	QUAN	CONC	VOL	QUAN	CONC	VOL	QUAN	CONC
Before	36	10,800	300	14.4	4320	300	21.6	6480	300
[a]Immed	+2	0	0	+2	0	0			
				16.4	4320	263			
[b]Ststate	38	10,800	284	15.2	4320	284	22.8	6480	284
[c]Vol Change	+1			+0.8			+1.2		

[a]Immed = immediate effect.   [b]Ststate = final steady-state.   [c]Vol Change = final volume change.

Absorption of the ingested water expands (+2 L) and dilutes the ECF (to 263 mOsm/L), creating an osmotic gradient between the ECF and the ICF. Water then diffuses from the less concentrated ECF (lower solute concentration but higher water concentration) to the ICF. As water leaves the ECF, the fluid in that compartment becomes more concentrated while the ICF becomes less concentrated. This process stops when the ECF and the ICF have the same osmolar concentration (284 mOsm/L). Then the osmolar concentration of all parts of the TBW are the same and can be calculated from its volume and osmole content: $C = Q/V$. The volume of the ECF and the ICF can be calculated using the same relationship and the fluid concentration calculated above, $V = Q/C$.

The final outcome is an expansion of the ECF (which contains 40% of the solute) by 0.8 L (40% of the volume absorbed) and of the ICF (which contains 60% of the solute) by 1.2 L (60% of the volume absorbed). Thus drinking water *expands all compartments in proportion to the number of osmoles in each, and it equally dilutes the fluid in all compartments.* See Figure co 2-1.

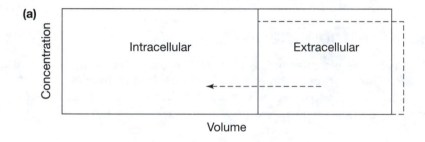

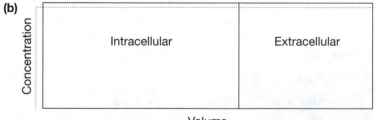

Figure co 2-1. Two Darrow-Yanet diagrams, plots of concentration versus volume of the intracellular and extracellular fluids. The solid lines show the normal status of the compartments. In (a), the effects of drinking water are shown (dashed lines). The ECF is expanded and diluted creating an osmotic gradient between the ECF and the ICF. Water will move in the direction shown by the arrow. The steady-state (before excretion of the water load) is shown in (b). Both compartments are diluted and expanded from their control states.

UNIT 10

Body Fluid Compartments 3

HYPEROSMOLAR FLUID EXPANSION

PHYSIOLOGICAL BACKGROUND

The Na-pump (Na/K-ATPase) extrudes Na from cells as rapidly as it enters. Hence, cell membranes are effectively impermeable to Na (and its associated anions), and all the solute in an NaCl infusion remains in the ECF. This problem deals with the effects of infusing a hyperosmotic saline solution on the volume and concentration of the body fluids.

See Compartments 2 for the method of calculating the volumes and osmolar contents of the fluid compartments of a man of average build.

PROBLEM INTRODUCTION

Gain or loss of water and/or electrolytes affects the volume, composition, and concentration of body fluids, and changes in these fluid compartment properties can have significant effects on body functions.

A 60-kg man of average build was rapidly infused with 0.5 L of 2.925% NaCl. Assume that he excretes neither salt nor water.

1. What would be the final volume and osmolarity of his fluid compartments? Show your calculations and enter your answers in the table below.

It is first necessary to determine how many osmoles of solute were administered:

$$2.925\% \text{ NaCl} = 2.925 \text{ gm/100 ml} = 29.25 \text{ gm/L}$$

The molecular weight of NaCl is 58.5, so the infused solution is 29.25/58.5 = 0.5 molar NaCl. Each NaCl dissociates into one Na^+ and one Cl^-, so the infused solution is

$$0.5 \text{ M} \times 2 \text{ Osm} = 1 \text{ Osm} = 1 \text{ osmole/L} = 1000 \text{ mOsm/L}$$

In 0.5 L, there would be $1000 \times 0.5 = 500$ mOsm of solute.

TIME	TOTAL BODY WATER			EXTRACELLULAR FLUID			INTRACELLULAR FLUID		
	VOL	QUAN	CONC	VOL	QUAN	CONC	VOL	QUAN	CONC
Before	36	10,800	300	14.4	4320	300	21.6	6480	300
aImmed	+0.5	1500		+0.5	+500				
				14.9	4820	323			
bStstate	36.5	11,300	309.6	15.57	4820	309.6	20.93	6480	309.6
cVOL Change	+0.5				+1.17		−0.67		

aImmed = immediate effect.   bStstate = final steady-state.   cVol Change = final volume change.
Note: See answer to Compartment 2 for calculations for the "Before" entries.

One can calculate the immediate effect of the saline administration on the concentration of the ECF from the new volume and the ECF compartment osmole content: $C = Q/V$. As expected, administering a hyperosmolar fluid concentrated his ECF (from 300 to 323 mOsm/L).

The increase in ECF concentration created an osmolar difference between the ECF and the ICF. Water then osmosed (diffused) from the less concentrated ICF to the more concentrated ECF until the concentrations were again equal. The final steady-state concentration can be determined from the total solute content and the volume of the TBW: $C = Q/V = 309.6$ mOsm/L.

One can therefore see that infusing a hyperosmotic solution increased the concentration of all the body fluid compartments. Because the infused solute was confined to the ECF, it created an osmotic gradient that caused water to leave the cells (0.67 L). Therefore, the volume of the ECF increased by the amount of fluid infused (0.5 L) plus the amount that moved from the ICF to the ECF (0.67 L plus 0.5 L = total ECF increase = 1.17 L). See Figure co 3-1.

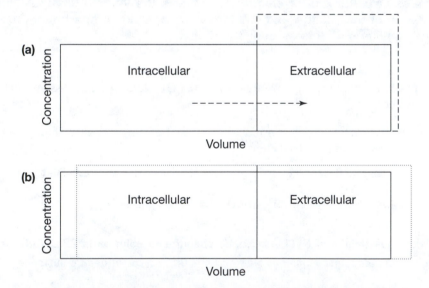

Figure co 3-1. Two Darrow-Yanet diagrams illustrate the effects of hyperosmotic saline infusion. The solid lines are the preinfusion values of volume and concentration. Part (a) shows the effects of hyperosmotic saline infusion. The extracellular fluid (ECF) volume and concentration are both increased (dashed lines). The osmolar concentration difference between the ECF and the intracellular fluid (ICF) causes water to move from the ICF to the ECF, as shown by the arrow. Part (b) shows the steady-state before any of the infusate is excreted (dotted lines). The concentration of both ECF and ICF are increased. The ICF volume is reduced, and the ECF volume is further increased.

2. How much water would he have to drink to restore the osmolarity of his body fluids to normal? (Assume no excretion of either salt or water.)

The original infusate was hyperosmolar and increased the concentration of his body fluids. To restore his body fluids to normal, he would have to drink enough water to dilute his total osmoles (11,300 mOsm) to 300 mOsm/L:

$$\frac{11,300 \text{ mOsm}}{300 \text{ mOsm/L}} = 37.67 \text{ L}$$

His present fluid volume is 36.5 L. Therefore, he must drink 37.67 L − 36.5 L = 1.17 L.

3. What would be the volume of his fluid compartments after he absorbed this water? Enter your answers in the table below.

TIME	TOTAL BODY WATER			EXTRACELLULAR FLUID			INTRACELLULAR FLUID		
	VOL	QUAN	CONC	VOL	QUAN	CONC	VOL	QUAN	CONC
Before	36.5	11,300	309.6	15.57	4820	309.6	20.93	6480	309.6
[a]Immed	+1.17	0		+1.17	0				
				16.74	4820				
[b]Stsate	37.67	11,300	300	16.06	4820	300	21.6	6480	300
[c]Vol Change				+1.67			0		

[a]Immed = immediate effect.   [b]Stsate = final steady-state.   [c]Vol Change = final volume change.

Because all the ICF solute remained in his cells, restoring the osmolarity of the body to normal returned the ICF volume to normal ($V = Q/C$). In the final steady state, all the water that was infused and all the water that he drank ended up in the ECF because the infused NaCl is impermeable to his cells. The water and solute are confined to the ECF until they are excreted. See Figure co 3-2.

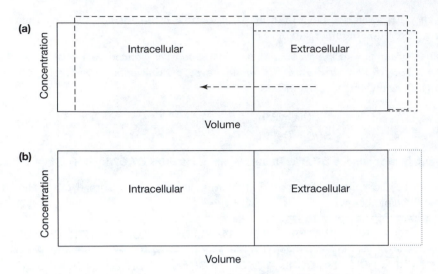

Figure co 3-2. Two Darrow-Yanet diagrams illustrate the effects of drinking enough water to restore body fluid concentrations to normal after a hyperosmotic saline infusion. In part (a), the long dashed lines show the steady-state after saline infusion, and the short dashed lines show the initial effects of drinking water. Extracellular fluid volume is increased, and the ECF is diluted. This process creates an osmotic gradient, causing water to move from the ECF to the more concentrated intracellular fluid (ICF). Part (b) shows the final steady state. Both ECF and ICF concentrations are once again normal. The ICF volume is also restored to normal, but the ECF volume is further expanded (dotted line).

Acid-Base Balance 1

ACID-BASE BALANCE

PHYSIOLOGICAL BACKGROUND

The [H^+] (or pH; pH $= -\log$ [H^+]) of the extracellular fluid is very closely regulated. This regulation involves extracellular and intracellular buffers and compensatory mechanisms. Buffers are passive chemical mechanisms for limiting the pH change that is caused by acid-base challenges. Compensatory mechanisms are the result of active physiological responses. Two compensatory mechanisms, renal and respiratory, exist. The three processes (buffering, respiratory compensation, and renal compensation) occur over different times. Buffering is very rapid, respiratory compensation is slower, and renal compensation is the slowest.

Respiratory compensation results from chemoreceptor-driven changes in alveolar ventilation. Alkalosis inhibits ventilation, whereas acidosis stimulates it. Therefore, in alkalosis, PCO_2 and [H^+] are raised, moving the pH toward normal. In an acidosis, PCO_2 and [H^+] are decreased.

The kidneys regulate pH by adjusting their H^+ secretion rate. In alkalotic situations, renal H^+ secretion decreases, thereby allowing part of the filtered HCO_3^- to be excreted. This process reduces the plasma [HCO_3^-], and the plasma becomes more acidic. Acidosis causes the kidneys to secrete more H^+. In the process, the kidneys manufacture new HCO_3^-, which is delivered to the plasma. This new HCO_3^- reacts with excess plasma H^+, restoring the pH towards or to normal.

The Davenport nomogram (Figure abase 1-2) is one of a number of commonly used acid-base representations. The figure is a plot of the Henderson-Hasselbalch relationship for the plasma carbonic acid buffer system. The pH is represented on the abscissa, with alkaline conditions to the right and acid conditions to the left. Plasma [HCO_3^-] is plotted on the ordinate, and values of constant PCO_2 are shown as isobars, with PCO_2 increasing to the left. The range of normal values is shown by the polygon in the center of the diagram; average normal values are shown by the intersection of the pH $= 7.4$ and [HCO_3^-] $= 24$ mM/L lines and the $PCO_2 = 40$ isobar.

Three other important features are shown. The Hb titration line defines the initial trajectory, followed by the variables when there is a primary change in PCO_2 (respiratory acidosis and alkalosis). It is also the slope of the trajectory followed during respiratory compensation. Line RC defines the maximum change that can be produced by respiratory compensation. The MKC region defines the maximum ability of the kidneys to increment the plasma [HCO_3^-] during renal compensation for a respiratory acidosis.

CLINICAL BACKGROUND

A wide variety of conditions affect acid-base balance: (1) changes in respiratory or renal function, (2) excess loss of GI secretions in vomiting and diarrhea, (3) certain endocrine disorders, and (4) excess production of intermediary metabolites as in anaerobic exercise or diabetic ketosis. Such acid-base imbalances complicate the problems that give rise to them and may make these conditions difficult to treat.

PROBLEM INTRODUCTION

The status of individuals with acid-base problems may be determined from the blood levels of the components of the carbonic acid buffer system. Plotting these values on an acid-base nomogram, such as the Davenport nomogram, helps one evaluate the individual's status. It also provides information that allows one to infer aspects of the individual's history and possible causes for the condition that exists.

The acid-base status of five individuals is given below. Plot the data for each individual on the Davenport nomogram (Figure abase 1-1).

INDIVIDUAL	pH	HCO_3^- (mM)	PCO_2 (mm Hg)
A	7.35	33.0	60
B	7.25	34.1	80
C	7.28	18.3	40
D	7.50	23	30
E	7.51	36	46

Data points for the five individuals are plotted on the Davenport nomogram in Figure abase 1-2.

Fill in the table below for the five individuals.

SUBJECT	A	B	C	D	E
Disturbance (resp/met, acid/alk[a])	RESPIRATORY; ACIDOSIS	RESPIRATORY; ACIDOSIS	METABOLIC; ACIDOSIS	RESPIRATORY; ALKALOSIS	METABOLIC; ALKALOSIS
Compensation (none/part/full, resp/kidney)	FULL; KIDNEY	PARTIAL; KIDNEY	NONE	NONE	FULL; RESPIRATORY
What will happen next?	NOTHING	MORE RENAL COMPENSATION	RESPIRATORY COMPENSATION	RENAL COMPENSATION	RENAL COMPENSATION
Base excess/deficit amount	EXCESS, 8.1 mM	EXCESS, 8.1 mM	DEFICIT, 7.8 mM	NONE	EXCESS, 13.4 mM
Possible cause	CO_2 RETENTION, E.G., EMPHYSEMA	CO_2 RETENTION, E.G., PNEUMONIA	NONCARBONIC ACID EXCESS, E.G., KIDNEY FAILURE	CO_2 DEFICIT, E.G., ALTITUDE	EXCESS BICARBONATE ACCUMULATION, E.G., VOMITING

a Resp/met = respiratory or metabolic; acid/alk = acidosis or alkalosis.

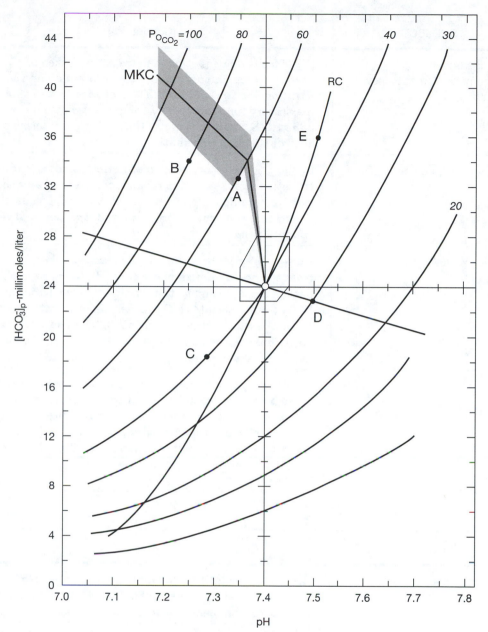

Figure abase 1-2. The data points of the five individuals in the problem are plotted in the Davenport nomogram. See the text for a discussion.

Explanation of the answers is given below.

SUBJECT	A
Disturbance (resp/met, acid/alk)	<u>Respiratory acidosis.</u> This condition is shown by the combination of an elevated PCO_2 (the cause) and a low pH (the consequence).

SUBJECT	A (Continued)
Compensation (none/part/full, resp/kidney)	**Full kidney compensation.** The trajectory taken in developing respiratory acidosis is leftward along the Hb titration line to whatever PCO_2 is caused by the abnormal respiratory function. Kidney compensation then slowly adds HCO_3^- to the plasma. The MKC area determines the limit of kidney compensation.
What will happen next?	**Nothing.** Respiratory compensation does not occur in conditions caused by the respiratory system, and the kidneys have done as much as they can. So, unless the respiratory problem is corrected, the individual's state cannot change.
Base excess/ deficiency, amount	**Excess, 8.1 mM.** This amount is measured directly from the nomogram as the vertical distance from the patient's data point to the Hb titration line (33.0 mM HCO_3^- − 224.9 mM HCO_3^-). Because the data point is above the Hb titration line, it is a base excess. The extra buffer base is the new HCO_3^- produced by the kidney during renal compensation.
Possible cause	**Emphysema.** A respiratory problem must be present for at least 3 days for complete renal compensation to be achieved. In fact, the cause could be any chronic condition that interfered with pulmonary CO_2 exchange (CO_2 retention): reduced ventilation, reduced diffusion capacity (e.g., pneumonia), or chronic lung disease (e.g., emphysema).

SUBJECT	B
Disturbance (resp/met, acid/alk)	**Respiratory acidosis.** The same criteria apply here as in individual A: increased PCO_2 and acid pH together are indicators of respiratory acidosis.
Compensation (none/part/full resp/kidney)	**Partial kidney.** The individual's history is also like A's, except that renal compensation has not reached its limit, the MKC area.
What will happen next?	**More renal compensation.** Given additional time, the kidneys could raise the $[HCO_3^-]$ to the MKC area. Meanwhile, the "sick" respiratory system would maintain the PCO_2 at its present value of 80.
Base excess/ deficiency, amount	**Excess, 8.1 mM.** Again, the value is an excess of 8.1 mM, determined from the nomogram. It is the vertical distance from the Hb titration line to the individual's acid-base data point (34.1 mM HCO_3^- − 26.0 mM HCO_3^- = 8.1 mM).

SUBJECT	B (Continued)
Possible cause	<u>Pneumonia</u>. The respiratory problem must be of less than about 3 days' duration. If the problem had been present for 3 days, renal compensation would be complete. One possibility is that this is subject A who has acutely contracted another respiratory problem, for example, a person with a chronic obstructive lung disease like emphysema who has <u>just</u> gotten pneumonia (no chance for additional renal compensation). In that case, the pneumonia would have worsened the original condition, and the trajectory that would have been taken would be parallel to the Hb titration line. Subjects A and B would then have the same base excess, as they do, until the kidneys can add more new HCO_3^-.

SUBJECT	C
Disturbance (resp/met, acid/alk)	<u>Metabolic acidosis</u>. The subject exhibits a low pH. PCO_2 is normal, however, and therefore cannot have been the cause of the acidosis. Further, plasma $[HCO_3^-]$ is reduced as would occur after HCO_3^- buffered an added noncarbonic acid load.
Compensation (none/part/full, resp/kidney)	<u>None</u>. Compensation by the respiratory system would be the first step to occur in response to the acidosis after buffering. During respiratory compensation, ventilation would increase, causing the PCO_2 to fall. No reduction in PCO_2 is present, however.
What will happen next?	<u>Respiratory compensation</u>. The excess H^+ will stimulate the peripheral chemoreceptors, causing an increase in alveolar ventilation and a consequent fall in PCO_2. The excess H^+ that is present will then be reduced.
Base excess/ deficiency, amount.	<u>Deficit, 7.8 mM</u>. The acid-base data point lies below the Hb titration line through the normal point. Therefore, the subject has a base deficit. The magnitude of the deficit is given by the vertical distance between the data point and the Hb line (25.6 mM HCO_3^- −18.3 mM HCO_3^- = 7.8 mM).
Possible cause	<u>Kidney failure</u>. Any condition that would produce an accumulation of noncarbonic acid could be the cause. Examples are renal failure (the kidney normally excretes the metabolically produced noncarbonic acids) or excess production of an acidic intermediary metabolite (e.g., lactic acid in anaerobic exercise or ketones in uncontrolled diabetic ketoacidosis).

SUBJECT	D
Disturbance (resp/met, acid/alk)	<u>Respiratory alkalosis.</u> The combination of a low PCO_2 (the cause) and an alkalosis (the result) defines this condition.
Compensation (none/part/full, resp/kidney)	<u>None.</u> The individual's acid-base data point lies on the Hb titration line, the initial trajectory taken by all primary respiratory disturbances.
What will happen next?	<u>Renal compensation.</u> Respiratory compensation cannot occur in respiratory acid base disorders. The respiratory system has caused the problem; and therefore it cannot repair it. Hence, renal compensation is the next event that will occur.
Base excess/ deficit, amount	<u>None.</u> The individual's acid-base data point lies on the Hb titration line. That is the line of constant buffer base content. Every point on the line represents a state in which the total body buffer base content is the same as in the normal state.
Possible cause	<u>High altitude.</u> Any condition that causes an increase in alveolar ventilation unaccompanied by a proportional increase in metabolism will decrease PCO_2. Examples are exposure to high altitude (the low PO_2 acts on the peripheral chemoreceptors and stimulates ventilation) and drug stimulation of the central nervous system (overdose of aspirin directly drives ventilation).

SUBJECT	E
Disturbance (resp/met, acid/alk)	<u>Metabolic alkalosis.</u> This person is alkalotic and has a change in PCO_2. The PCO_2, however, is increased from normal and therefore cannot have been the cause of the alkalosis. An increased PCO_2 would cause an acidosis. This is a metabolic disturbance.
Compensation (none/part/full, resp/kidney)	<u>Full, respiratory.</u> The person's acid base data point lies on the maximum respiratory compensation line; that is, the increase in plasma PCO_2 is the result of respiratory compensation.
What will happen next?	<u>Renal compensation.</u> Renal compensation is slow. It occurs after respiratory compensation.
Base excess/ deficit, amount	<u>Excess, 13.4 mM.</u> The acid-base data point is above the Hb titration line by that amount (36 mM HCO_3^- − 22.6 mM HCO_3^- = 13.4 mM).
Possible cause	<u>Vomiting.</u> Vomiting (the HCO_3^- that is produced during acid secretion by the stomach accumulates in the body) or the ingestion of alkaline materials (antacid tablets) will cause a metabolic alkalosis.

Acid-Base Balance 2

DIABETIC KETOACIDOSIS

PHYSIOLOGICAL BACKGROUND

Normally, acids that are produced metabolically are excreted by way of the lungs (as CO_2) or in the urine (phosphates and sulfates). Excretion occurs as rapidly as production. In that way, the arterial pH is not affected by acid production. Under conditions in which the rate of gain or loss of acidic or basic materials increases, however, the arterial pH does change. In those instances, the mechanisms available for the regulation of pH come into play to limit the effect of the imbalance and to restore the pH toward or to its normal value. This restoration is essential because changes in pH alter the function of essential protein materials (e.g., enzymes, receptors).

Three mechanisms—buffering, respiratory compensation, and renal compensation—guard against excessive changes in pH and restore the pH. Buffering is the first line of defense. In this passive chemical process, weak bases and/or weak acids react with excess H^+ or liberate H^+. They thereby reduce the change in pH that occurs when acid is gained or lost by the body. Buffering is very rapid, but it is only a temporary expedient. There is a limited amount of buffer in the body, and the process of buffering uses it up. Hence, the buffers must be ultimately restored.

Compensation is an active physiological process that restores the buffers that have reacted and returns the pH toward or to normal. There are two compensatory mechanisms, respiratory and renal. Respiratory compensation is the faster of the two. It depends on a change in the activity of the peripheral chemoreceptors caused by altered arterial pH. The chemoreceptors then cause a reflex change in alveolar ventilation, resulting in an appropriate modification of arterial PCO_2 (a decrease in acidosis and an increase in alkalosis). This compensatory process shifts the pH toward normal; it cannot, however, completely return the pH to a normal value. In metabolic acidosis, for example, the low pH stimulates ventilation. The resulting fall in PCO_2, however, partially removes a tonic stimulus to the central chemoreceptor. The balance between the stimulating and inhibiting effects yields an alveolar ventilatory rate that is insufficient to completely correct the pH. Maximum possible respiratory compensation is shown by the line RC in the Davenport acid base nomogram shown in Figure abase 2-1.

Renal compensation results from the reaction of the renal tubule cells to a change in pH. In acidosis, there is an increase in H^+ secretion and ammonium production. Both processes lead to an increased HCO_3^- synthesis. This HCO_3^- is transferred to the body, where it reacts with and neutralizes excess H^+. In alkalosis, by contrast, the kidneys excrete part of the filtered HCO_3^- and may secrete additional HCO_3^-. The loss of HCO_3^- from the body shifts the reaction of CO_2 and water, leading to the production of compensating H^+.

The two compensatory mechanisms take place over different time intervals. Respiratory compensation is about 95% complete in 12 hours. Renal compensation takes 1 day (alkalosis) to 3 days (acidosis) to reach completion.

CLINICAL BACKGROUND

Diabetes mellitus is a common endocrine disorder (see Endocrine 3). One consequence of uncontrolled, insulin-dependent diabetes is an increase in fat mobilization and in the conversion of the resulting fatty acids to acetoacetic acid and β-hydroxybutyric acid, the so-called ketones. The ketones are extremely good energy-supplying substrates, but, when produced at a rate that exceeds their consumption, they cause a metabolic acidosis to develop.

One consequence of this acidosis is ventilatory stimulation, which produces a characteristic ventilatory pattern called Kussmaul's breathing. This, of course, is the respiratory response that results in respiratory compensation for the metabolic acidosis that is present.

PROBLEM INTRODUCTION

Acid-base homeostasis is essential for normal function because changes in pH affect all proteins (enzymes, receptors, etc.). Hence, rapidly acting mechanisms exist that buffer any pH changes that might occur as a result of excess acid or base accumulation or loss. In addition, powerful compensatory mechanisms exist that permanently, but more slowly, correct any deviations from normal pH.

Jenny is an insulin-dependent diabetic (see Endocrine 3). She was admitted to the hospital in a coma with ketonuria because she had not taken her insulin. Ketonuria occurs when the plasma ketone level rises high enough (ketonemia) for the filtered ketone load to exceed the renal threshold. Ketones are acids.

1. Predict the changes from her normal values (increase, decrease, no change) that you would expect her to exhibit as a result of her ketonemia.

VARIABLE	CHANGE
arterial pH	DECREASE
arterial [HCO_3^-]	DECREASE
arterial PCO_2	NO CHANGE
noncarbonic buffer base[a]	DECREASE
total base[b]	DECREASE

[a]Noncarbonic buffer base = total body content of all noncarbonic acid conjugate base.

[b]Total base = total base content of the body: bicarbonate + noncarbonic acid conjugate base.

2. Explain the cause of these changes.

Arterial pH

Adding ketones to the blood acidifies it. The pH change, however, is smaller than it might have been because much of the added H+ is buffered by the physiological buffer systems.

Arterial [HCO_3^-]

The carbonic acid buffer system circulates in the blood and is in the interstitial fluid. Hence, of the buffer systems, it is the most accessible to the excess H+ that is being produced. It consists of a weak acid (carbonic acid) and its conjugate base (HCO_3^-). Carbon dioxide, however, is usually used in place of carbonic acid in calculations because carbonic acid derives from CO_2 ($CO_2 + H_2O \longleftrightarrow H_2CO_3$). The added H+ is rapidly buffered by this system. The reactions shift to the right:

$$H^+ + HCO_3^- \longleftrightarrow H_2CO_3 \longleftrightarrow CO_2 + H_2O.$$

Thus $[HCO_3^-]$ is reduced in the process.

Arterial PCO_2

The ketones are added to the blood by the liver. Hence, the reaction shown above takes place in the hepatic sinusoids and the hepatic venous system. That blood is carried to the right heart and the lungs, where the elevated pulmonary capillary PCO_2 creates an increased partial pressure gradient, facilitating the diffusive loss of CO_2. The blood exiting the lungs has a PCO_2 that is essentially unchanged from normal. This blood returns to the heart and is pumped out as the systemic blood flow.

Conjugate base of the noncarbonic acid buffer systems

The noncarbonic (NC) acid buffers (hemoglobin, cellular proteins, and phosphates) are mostly located intracellularly. Their buffering action is delayed by the time required to transfer H^+ into the cells. Only hemoglobin—which, like the plasma component of the carbonic acid buffer system, circulates with the blood in the red blood cells—reacts in a short time (less than 1 hour). The others require much longer to exert their buffering action. When Hb^- (or any of the other NC bases), however, buffers H^+ ($H^+ + Hb^- \longleftrightarrow HHb$), the quantity of NC base is reduced.

Total base

The total body base content is a measure of how much conjugate base remains to buffer additional H^+ should there be another acid challenge. It is the sum of $[HCO_3^-]$ (the carbonic acid buffer base) and the NC acid buffer bases. Because these components have both been reduced by buffering the H^+ that dissociated from the ketones, total body base content has been reduced.

3. What is the first compensatory response that would occur (chemical buffering is not a compensatory response) to help restore her pH to normal?

   Respiratory compensation is the first to occur.

4. What is the stimulus for that compensation, and on what receptor system does it act?

   Respiratory compensation is brought about by an increase in ventilation due to peripheral chemoreceptor stimulation. The stimulus for this reflex is the increased ketone derived H^+ acting directly on the receptors.

The first compensatory response to occur is respiratory compensation.

5. Predict the changes (increase, decrease, no change) that this compensation would produce from her state *after the onset of ketonemia*

VARIABLE	CHANGE
arterial pH	INCREASE
arterial $[HCO_3^-]$	DECREASE
arterial PCO_2	DECREASE
noncarbonic buffer base [a]	INCREASE
total base [b]	NO CHANGE

[a] Noncarbonic buffer base = total body content of all noncarbonic acid conjugate base.

[b] Total base = total base content of the body, bicarbonate + noncarbonic acid conjugate base.

6. Explain your predictions.

Arterial PCO_2

Increased alveolar ventilation reduces the alveolar and hence the arterial PCO_2.

Arterial pH and $[HCO_3{}^-]$

A fall in arterial PCO_2 shifts the reactions to the right.

$$H^+ + HCO_3{}^- \longleftrightarrow H_2CO_3 \longleftrightarrow CO_2 + H_2O$$

This shift reduces both the $[H^+]$ (i.e., increases the pH) and the $[HCO_3{}^-]$.

Noncarbonic acid base content

When the $[H^+]$ is reduced, acid forms of the noncarbonic acid buffers (e.g., HHb) dissociate, increasing the quantity of the noncarbonic acid conjugate bases (e.g., Hb^-). For hemoglobin, the reaction is

$$HHb \longleftrightarrow H^+ + Hb^-$$

Total base

Total base is the sum of $HCO_3{}^-$ and the noncarbonic acid conjugate base contents. During respiratory compensation, this total does not change because the H^+ with which each $HCO_3{}^-$ reacts comes from the dissociation of noncarbonic buffer acid. Each $HCO_3{}^-$ that disappears is replaced by a noncarbonic acid base (e.g., Hb^-), virtually one for one. See the two equations above.

7. Predict the changes *from normal* in her acid-base variables (increase, decrease, no change) at the completion of respiratory compensation (about 95% complete in 12 hours).

VARIABLE	CHANGE
arterial pH	DECREASED
arterial $[HCO_3{}^-]$	DECREASED
arterial PCO_2	DECREASED
noncarbonic buffer base[a]	DECREASED
total base[b]	DECREASED

[a] Noncarbonic buffer base = total body content of all noncarbonic acid conjugate base.
[b] Total base = total base content of the body: bicarbonate + noncarbonic acid conjugate base.

8. Why does respiratory compensation not completely restore Jenny's arterial pH to normal?

Before respiratory compensation begins to correct Jenny's ketoacidosis, she has two main stimuli to ventilation: (1) her newly acquired increase in H^+, which will stimulate her peripheral chemoreceptors, and (2) her normal, tonic ventilatory stimulus, acting primarily through her central chemoreceptors. The second of these results from the normally acidic pH (7.3) of the cerebrospinal fluid (CSF).

When she begins to hyperventilate, her arterial PCO_2 starts to fall and correct her ketoacidosis. As her arterial PCO_2 falls, however, a PCO_2 gradient is created between the CSF and her circulation. As a result, CO_2 diffuses out of her CSF, lowering her CSF $[H^+]$, and removing part of her normal tonic ventilatory stimulation.

Hence, two opposing effects develop: an increased ventilatory stimulation due to increased H^+ acting on the peripheral chemoreceptors, and a decreased ventilatory stimulation due to decreased H^+ acting on the central chemoreceptors. The equilibrium point (i.e., the maximum possible respiratory pH compensation) is therefore somewhere between the pH that exists at the onset of respiratory compensation and the normal pH of 7.4. This balance is represented by the line RC in the Davenport nomogram shown in Figure abase 2-2.

Of course, in the process of respiratory compensation, additional HCO_3^- is lost along with H^+, additional CO_2 is expired, and some of the noncarbonic acid buffer forms dissociate to partially restore the NC base that had been consumed in the buffering process.

About the time of completion of respiratory compensation, Jenny was treated with insulin.

9. What is the next acid-base compensatory process that would occur?

   Renal compensation occurs next.

10. Predict the changes in her acid-base variables (increase, decrease, no change) that would occur during renal compensation.

VARIABLE	CHANGE
arterial pH	INCREASE
arterial [HCO_3^-]	INCREASE
arterial PCO_2	INCREASE
noncarbonic buffer base [a]	INCREASE
total base [b]	INCREASE

[a] Noncarbonic buffer base = total body content of all noncarbonic acid conjugate base.

[b] Total base = total base content of the body: bicarbonate + noncarbonic acid conjugate base.

11. What renal processes give rise to the increase in her plasma [HCO_3^-] during compensation?

    Two mechanisms give rise to the production of *new* bicarbonate by the kidneys: (1) the production of titratable acidity and (2) the production of ammonium ion. In producing titratable acidity, the renal tubule cells in the distal tubule and collecting ducts secrete H^+, which reacts with filtered (nonbicarbonate) anions in the tubular lumen. In the process of making the H^+ for secretion, a bicarbonate is produced and secreted into the body, as shown in Figure abase 2-2. When the urine is titrated to pH 7.4 with a strong base, the secreted H^+ that had reacted with the urinary base is removed (titratable acid), providing a measure of the amount of new HCO_3^- that was produced by this mechanism.

    The second mechanism by which the kidneys produce new bicarbonate involves the conversion of glutamine into ammonium ion and HCO_3^-. Both mechanisms are pH sensitive, increasing HCO_3^- production as the pH falls.

12. Compared with normal, what is the value of arterial pH at the completion of renal compensation (increased, decreased, normal)?

    Given sufficient time, renal compensation is able to return the arterial pH to normal in all conditions except respiratory acidosis and acid-base problems that are caused by the kidney. Hence, Jenny's pH change will be completely corrected.

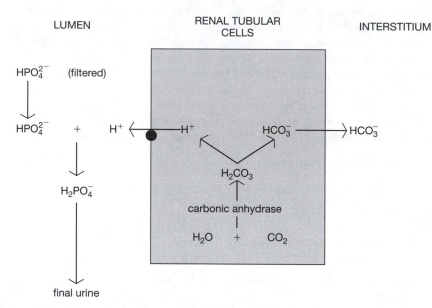

Figure abase 2-1. The production of titratable acidity by the kidney. (From A.J, Vander, *Renal Physiology*, 5th ed., McGraw-Hill, Inc. New York, 1995.)

13. What is the mechanism for the change in her arterial PCO_2 during renal compensation?

   The HCO_3^- that is secreted into the body by the kidney neutralizes excess H^+. As it does so, and as the $[H^+]$ declines, stimulation of the peripheral chemoreceptors decreases, reducing the loss of CO_2.

14. Plot points on the Davenport acid-base nomogram shown in Figure abase 2-2 that could reasonably represent Jenny's acid-base status:

   a. When she is normal

   b. As a result of ketonemia but before compensation

   c. At the completion of respiratory compensation

   d. Partway through renal compensation

   e. At the completion of renal compensation

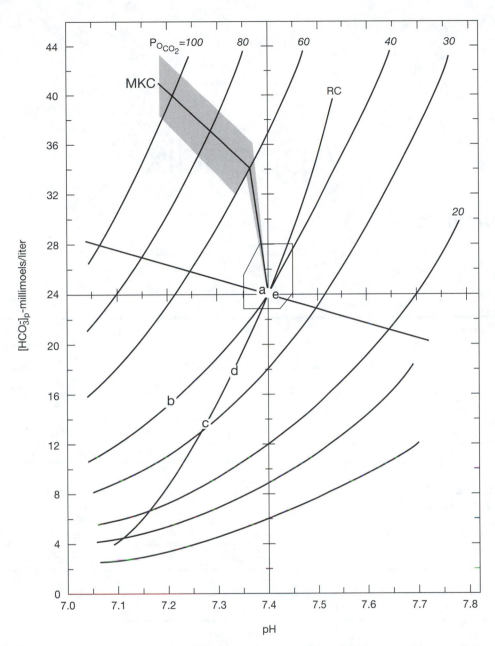

Figure abase 2-2. A Davenport acid-base nomogram showing the sequence of events in Jenny's acidosis. Point a is her normal state. Point b is her condition after the onset of ketoacidosis. Point c shows her acid-base status after respiratory compensation. Point d is partway through and point e is after complete renal compensation.

Appendix

Normal Values[1]

	Blood p = plasma s = serum	Urine
pH	7.38–7.44 (s)	4.6–8.0
osmolarity	280–295 (s) mOsm/L	500–800 mOsm/L
hematocrit-male	0.40–0.54	
hematocrit-female	0.37–0.47	
hemoglobin-male	14.0–18.3 gm/dl	
hemoglobin-female	11.8–17.5 gm/dl	
blood volume	60–80 ml/Kg	
Electrolytes		losses per 24 hours
bicarbonate	21–28 mEq/L (p)	highly variable
calcium	4.5–5.3 mEq/L (s)	6.5–16.5 m Eq
chloride	100–108 mEq/L (s)	120–240 mEq
potassium	3.8–5.0 mEq/L (p)	35–80 mEq
sodium	136–142 mEq/L (p)	120–220 mEq
Metabolites		
amino acids	2.3–5.0 mg/dl (p/s)	41–133 mg
creatinine	0.6–1.2 mg/dl (p/s)	1.01–2.50 mg
glucose	70–110 mg/dl	16–132 mg
Proteins		
total	6.0–7.8 gm/dl (s)	47–76.2 mg
albumin	3.2–4.5 gm/dl (s)	10–100 mg

[1] Modified from Martini, *Fundamentals of Anatomy and Physiology*, 4th edition.
Upper Saddle River, NJ: Prentice Hall, 1998.

Glossary

absolute refractory period A period of time immediately following an action potential during which a second action potential cannot be triggered, no matter how large the stimulus

absorption Transfer of substances from the lumen of the kidney or gastrointestinal tract to the extracellular space

absorptive state (fed state) The time following a meal when the products of digestion are being absorbed, used, and stored

acetylcholine (ACh) Neurotransmitter used by neurons of the parasympathetic nervous system

acetylcholinesterase Enzyme that breaks down acetylcholine in the synapse

acid A molecule that ionizes and contributes an H+ to a solution

actin A globular protein (G-actin) that polymerizes to form thin filaments (F-actin)

adenosine triphosphate (ATP) An energy-storing compound composed of adenine, ribose, and three phosphate group

adenylate cyclase Membrane-bound enzyme that converts ATP to cyclic AMP

adrenal cortex Outer portion of adrenal gland which produces steroid hormones

adrenal medulla Modified sympathetic ganglion, the inner portion of the adrenal gland that produces catecholamines

adrenergic Adjective pertaining to epinephrine (adrenaline) or norepinephrine

adrenocorticotropic hormone (ACTH) Anterior pituitary hormone that regulates secretion of cortisol from the adrenal cortex

agonist(s) Molecules that combine with a receptor and mimic the response of the natural reactant

aldosterone A steroid hormone that stimulates Na+ reabsorption and K+ secretion in the kidney

alpha motor neuron(s) Neurons that innervate skeletal extrafusal muscle fibers and cause muscle contraction

alveolar ventilation The volume of air that reaches the alveoli each minute

alveoli The smallest airways within the exchange surface of the lungs, where oxygen and carbon dioxide transfer between air and the blood

anabolism Metabolic pathways that require a net input of energy and that synthesize small molecules into larger ones

anaerobic Adjective pertaining to a process that does not require oxygen

androgen Steroid hormone produced in the gonads and adrenal cortex; dominant sex hormone in males

angiotensin II (AGII) Trophic hormone that regulates aldosterone secretion; also raises blood pressure and causes thirst and ADH secretion

angiotensin converting enzyme (ACE) Membrane-bound endothelial enzyme that converts AGI into AGII

antagonism One substance opposes the action of another, 150 of autonomic neurotransmitters and receptors

anterior pituitary An endocrine gland in the brain that secretes multiple hormones

antidiuretic hormone (ADH, vasopressin) Posterior pituitary hormone that regulates water reabsorption in the kidney

antiport Movement of two or more molecules in opposite directions across a membrane by a single transporter

aorta The main artery taking blood from the left ventricle to the body

aortic valve The valve between the left ventricle and the aorta

apical membrane The surface of transporting epithelial cells that face the lumen of an organ

arteriole(s) The smallest arteries and site of variable resistance in the circulatory system

artery(ies) Blood vessels that carry blood away from the heart

atrial natriuretic factor Peptide hormone from atria of the heart that increases renal Na+ and water excretion

atrioventricular node (AV node) The electrical gateway from the atria to the ventricles, located near the floor of the right atrium

atrium Upper chamber of the heart that receives blood from the blood vessels

autonomic neuron(s) Efferent neurons that control smooth muscle, cardiac muscle, many glands, and some adipose tissue

axon An extension of a neuron that carries signals to the target cell

axon terminal The distal end of a neuron where neurotransmitter is released into a synapse

baroreceptor(s) Stretch-sensitive mechanoreceptors that respond to changes in pressure

baroreceptor reflex The primary reflex pathway for homeostatic control of blood pressure

base (1) A molecule that decreases the H+ concentration of a solution by combining with free H+; (2) A carbon-nitrogen molecule with a ring structure that is an essential component of a nucleotide

basolateral membrane The sides of transporting epithelial cells that face the extracellular fluid

beta cells of the pancreas Endocrine cells that secrete insulin

blood-brain barrier Tight junctions in the brain capillaries that prevent free exchange of many substances between the blood and the cerebrospinal fluid

bronchiole(s) Small collapsible airways with smooth muscle walls

brush border Name given to the microvilli on some intestinal and renal epithelium

buffer A molecule that minimizes changes in pH

bulk flow Mass movement of water or air as the result of pressure gradients

bundle of His Specialized electrical conducting cells of the heart that carry signals into the ventricles

calcium-induced Ca2+ release Process in which Ca2+ entry into a muscle fiber triggers release of additional Ca2+ from the sarcoplasmic reticulum

capillary(ies) Smallest blood vessels where blood exchanges material with the interstitial fluid

carbaminohemoglobin Hemoglobin with bound carbon dioxide

carbonic anhydrase Enzyme that reversibly catalyzes the conversion of carbon dioxide and water into carbonic acid

cardiac cycle The period of time from the end of one heart beat through the end of the next beat

cardiac output The amount of blood pumped per ventricle per unit time

carrier protein Membrane transporter that binds to the molecule(s) it transports

catabolism Reactions that release energy and result in the breakdown of large biomolecules

catecholamine A type of signal molecule formed from tyrosine; includes epinephrine, norepinephrine, and dopamine

central nervous system (CNS) Integrating center for nervous reflexes composed of brain and spinal cord

central receptor(s) Sensory receptors located in or closely linked to the brain

cerebrospinal fluid (CSF) A salty solution that is continuously secreted into and absorbed from the ventricles of the brain

channel protein A membrane protein that forms water-filled channels to link intracellular and extracellular compartments

chemoreceptor A sensory receptor that is activated by binding of a chemical substance

chloride shift Process in which red blood cells transport one HCO_3^- in exchange for one Cl^- ion across the cell membrane

cholecystokinin (CCK) Intestinal hormone that regulates digestive function and may play a role in appetite

choroid plexus A transporting epithelium that secretes cerebrospinal fluid

chylomicron(s) Large droplets of triglycerides, cholesterol, proteins, and lipoproteins that are synthesized in cells of the small intestine

chyme A soupy substance produced by digestion in the digestive tract

cilia Short, hair-like structures whose movement creates currents that move fluids or secretions across the cell surface

colloid osmotic pressure (π) Osmotic pressure that is due to the presence of plasma proteins that cannot cross the capillary endothelium

competitive inhibitor(s) Molecules that bind to the active site of the enzyme, preventing substrate binding

concentration gradient A difference in the concentration of a substance between two places

conduction The movement of an action potential along the axon at high speed

connexin(s) Membrane-spanning proteins that form gap junctions; capable of opening and closing

contractility The intrinsic ability of a cardiac muscle fiber to contract at any given fiber length

coronary artery Artery supplying blood to the heart muscle

corpus luteum Ovarian structure that produces estrogen and progesterone after ovulation

cortex Literally, bark; the outer or surface portion of an organ

cortisol Steroid hormone from the adrenal cortex that regulates metabolism, particularly during stress

cotransporter A protein that moves more than one kind of molecule at one time

covalent bond(s) Bonds created by two atoms that share one or more pairs of electrons

creatinine The breakdown product of phosphocreatine

crossbridge Connection formed when mobile myosin heads bind to actin molecules in muscle

cyclic AMP (cAMP; cyclic adenosine-3', 5'-monophosphate) Nucleotide that participates in the transfer of signals between the external environment and the cell

cytoplasm All material inside the cell membrane except for the nucleus

cytosol Semi-gelatinous intracellular fluid containing dissolved nutrients, ions, and waste products

Dalton's Law The total pressure of a mixture of gases is determined by the sum of the pressures of the individual gases

dead space Those portions of the respiratory system that do not exchange gases with the blood

deamination Removal of an amino group from a molecule

dendrite(s) Thin, branched processes that receive and transfer incoming information to an integrating region within the neuron

deoxyribonucleic acid (DNA) Nucleotide that stores genetic information in the nucleus

depolarization A decrease in the membrane potential difference of a cell

dextrose A six-carbon sugar; also known as glucose

diastole The time during which cardiac muscle is relaxed

diastolic pressure Lowest pressure in the arteries associated with relaxation of the ventricles

diffusion Movement of molecules from an area of higher concentration to an area of lower concentration

digestion Chemical and mechanical breakdown of foods into smaller units that can be absorbed

2,3-diphosphoglycerate (2,3-DPG) A metabolite of red blood cells that lowers the binding affinity of hemoglobin for oxygen

down-regulation Decrease in receptor number or binding affinity that lessens the response of the target cell

edema The accumulation of fluid in the interstitial space

effector The cell or tissue that carries out the homeostatic response

efferent neuron A peripheral neuron that carries signals from the central nervous system to the target cell

eicosanoid(s) Modified 20-carbon fatty acids that act as regulators of physiological funtions

electrocardiogram (ECG, EKG) A recording of the electrical activity of the heart

electrochemical gradient The combined concentration and electrical gradient for an ion

end-diastolic volume (EDV) The maximum volume of blood that the ventricles hold during a cardiac cycle

endocrine gland A ductless gland or single cell that secretes a hormone

endocytosis Process by which a cell brings molecules into the cytoplasm in vesicles formed from the cell membrane

endoplasmic reticulum A network of interconnected membrane tubes in the cytoplasm; site of protein and lipid synthesis and calcium storage

endothelin(s) A family of powerful vasoconstrictors released by the vascular endothelium

endothelium Layer of thin epithelial cells that line the lumen of the heart and blood vesssels

endothelium-derived relaxing factor (EDRF) Nitric oxide released by endothelial cells; relaxes vascular smooth muscle

end-plate potential (EPP) Depolarization at the motor end plate due to acetylcholine

end-systolic volume (ESV) The amount of blood left in the ventricle at the end of contraction

epiphyseal plate Region of long bones where active longitudinal bone growth takes place

epithelium Tissue that protects surface of the body, lines hollow organs, and manufactures and secretes substances

equivalent (eq) Molarity of an ion times the number of charges the ion carries

erythrocyte(s) Red blood cells that transport oxygen and carbon dioxide between the lungs and the tissues

erythropoietin (EPO) Hormone made in the kidneys that regulates red blood cell production

essential amino acid(s) Nine amino acids the human body cannot synthesize and must obtain from the diet

estrogen Steroid hormone produced in ovary and adrenal cortex; dominant steroid in females

excitable tissue Neural and muscle tissue that is capable of generating and responding to electrical signals

excitation-contraction coupling The sequence of action potentials and Ca^{2+} release that initiate contraction

excitatory postsynaptic potential(s) (EPSPs) Depolarizing graded potentials that make a neuron more likely to fire an action potential

excretion The elimination of material from the body at the lungs, intestine, or kidneys

exocrine gland A gland that releases secretions into the external environment through ducts

exocytosis Process in which intracellular vesicles fuse with the cell membrane and release their contents into the extracellular fluid

expiration The movement of air out of the lungs

external environment The environment sur-rounding the body

external respiration The interchange of gases between the environment and the body's cells

extracellular fluid The internal fluid that surrounds the cells

facilitated diffusion Movement of molecules across cell membranes in response to a concentration gradient with the aid of a membrane protein

feedback loop Information about a homeostatic response that is sent back to the integrating center

feedforward control Anticipatory responsees that start a response loop in anticipation of a change that is about to occur

fibrin Protein polymer fibers that stabilize platelet plugs

filtration fraction The percentage of the renal plasma flow that filters at the glomerulus

first messenger Chemical signal molecules released by cells

flow rate The volume of blood that passes one point in the system per unit time

fluid mosaic model Membrane composed of phospholipid bilayer with proteins inserted wholly or partially into the bilayer

fluid pressure Pressure created by the presence of fluid within an enclosed space

follicle (1) Group of cells in the ovary that contains a developing egg; (2) Group of thyroid cells that produce thyroid hormones

follicle stimulating hormone (FSH) Anterior pituitary hormone that stimulates gamete production in the gonads

functional unit The smallest structure that can carry out all the functions of an organ

G protein(s) Membrane proteins that link membrane receptors to either ion channels or membrane enzymes

ganglion A cluster of nerve cell bodies in the peripheral nervous system

gap junction(s) Cytoplasmic bridges between adjacent cells, created by linked membrane proteins

gastrin Hormone secreted by the stomach that stimulates gastric acid secretion

gland Group of epithelial cells specialized for synthesis and secretion of substances

glomerular filtration Movement of fluid from the glomerulus into the lumen of Bowman's capsule

glomerular filtration rate (GFR) The amount of fluid that filters into Bowman's capsule per unit time

glucagon Pancreatic hormone that elevates plasma glucose

glucocorticoids Adrenal steroid hormones such as cortisol that elevate plasma glucose

gluconeogenesis Pathways through which non-carbohydrate precursors, especially amino acids, are converted into glucose

glucosuria (glycosuria) Excretion of glucose in the urine

glycerol A simple 3-carbon molecule that is the backbone of fatty acids

glycogen Storage polysaccharide found in animal cells

glycogenesis The synthesis of glycogen from glucose

glycogenolysis The breakdown of glycogen

glycolysis Metabolic pathway that converts glucose to pyruvate (aerobic) or lactic acid (anaerobic)

gonadotropin (FSH and LH) A type of peptide hormone from the anterior pituitary that acts on the gonads

gonadotropin releasing hormone (GnRH) Hypothalamic hormone that stimulates release of gonadotropins from the anterior pituitary

graded potential A change in membrane potential whose magnitude is proportional to the stimulus and that decreases with distance as it spreads throught the cytoplasm

growth hormone A peptide hormone from the anterior pituitary that controls tissue growth

half-life The amount of time required to reduce the concentration of hormone by one-half

hematocrit Percentage of the total blood volume that is packed red blood cells

hematopoiesis Blood cell production in the bone marrow

hemoglobin Oxygen-carrying pigment of red blood cells

hemostasis Process of keeping blood within the blood vessels by repairing breaks without compromising the fluidity of the blood

hepatic portal system Specialized region of the circulation that transports material absorbed at the intestine directly to cells of the liver

histamine Paracrine secreted by mast cells and basophils; acts as a vasodilator and broncho-constrictor

homeostasis The ability of the body to maintain a relatively constant internal environment

hormone A chemical secreted by a cell or group of cells into the blood for transport to a distant target where it acts in very low concentrations to affect growth, development, homeostasis, or metabloism

hydraulic pressure Pressure exerted by fluid

hydrogen bond(s) Weak attractive forces between hydrogens and other atoms, especially oxygen and nitrogen

hydrolysis Reaction in which large molecules are broken into smaller ones by addition of water

hydrophilic molecule Molecules that dissolve readily in water

hydrophobic molecule Molecules that do not dissolve readily in water

hydrostatic pressure The pressure exerted by a stationary column of fluid

hypercapnia Elevated P_{CO_2} in the blood

hyperplasia Increased cell number due to cell division

hyperpolarization A membrane potential that is more negative than the resting potential

hypertension Elevated blood pressure

hypertonic solution A solution that causes net movement of water out of a cell

hypertrophy An increase in cell size without an increase in cell number

hyperventilation An increase in alveolar ventilation that is proportionately more than an increase in metabolic rate

hypothalamus Region of the brain that contains centers for behavioral drives and plays a key role in homeostasis

hypotonic solution A solution that causes a net influx of water into a cell

independent variable The parameter manipulated by the investigator in an experiment

inhibitory postsynaptic potential(s) (IPSP) Hyperpolarizing graded potentials that make a neuron less likely to fire an action potential

inotropic agent Any chemical that affects contractility

inspiration The movement of air into the lungs

inspiratory muscles Muscles of inspiration, including external intercostals, diaphragm, scalenes, and sternocleidomastoids

insulin Pancreatic hormone that decreases plasma glucose concentration

integrating center The control center that evaluates incoming signal and decides on an appropriate response

intercalated disk(s) Specialized cell junctions in cardiac muscle that contain gap junctions

intercostal muscles Muscles associated with the rib cage; used for breathing

internal environment The extracellular fluid that surrounds the cells of the body

interstitial fluid Extracellular fluid that surrounds the cells and lies between the cells and the plasma

intrapleural pressure Pressure within the pleural fluid

intrapulmonary pressure Pressure within air spaces of lungs

inulin A polysaccharide isolated from plants; used to determine extracellular fluid volume and glomerular filtration rate

ion An atom with a net positive or negative charge due to gain or loss of one or more electrons

ionic bond A bond between ions attracted to each other by opposite charge

islets of Langerhans Clusters of endocrine tissue within the pancreas

isometric contraction A contraction that creates force without movement

isotonic contractraction A contraction that creates a constant force and moves a load

isotonic solution A solution that results in no net water movement when a cell is placed in it

juxtaglomerular apparatus Region where the distal tubule of the nephron passes between afferent and efferent arterioles

juxtaglomerular (JG) cell(s) Specialized cells in the walls of renal arterioles that synthesize and release renin

ketoacidosis A state of acidosis that results from excessive ketone production

kilocalorie (kcal or Calorie) Amount of energy needed to raise the temperature of 1 liter of water by $1°$ C

kinase An enzyme that adds a phosphate group to the substrate

lacteal A fingerlike projection of the lymph system that extends into the villi of the intestine

lactic acid The endproduct of anaerobic glycolysis

latent period Delay between the muscle action potential and beginning of muscle tension that represents the time required for Ca^{2+} release and binding to troponin

law of mass action For a reaction at equilibrium, the ratio of substrates to products is always the same

law of mass balance If the amount of a substance in the body remains constant, any gain must be offset by an equal loss

Leydig cell(s) Testicular cells that secrete testosterone

liposome(s) Spherical structures with an exterior composed of a phospholipid bilayer, leaving a hollow center with an aqueous core

local control Homeostatic control that takes place strictly at the tissue or cell by using paracrine or autocrine signals

lung capacity(ies) Sums of two or more lung volumes

luteal phase The portion of the menstrual cycle following ovulation, when the corpus luteum produces strogen and progesterone

luteinizing hormone (LH) Anterior pituitary hormone that acts on the gonads to influence hormone production

lymph The fluid within the lymphatic system that moves from the tissues to the venous side of the systemic circulation

macrophage(s) Tissue phagocytes that develop from monocytes

macula densa Specialized cells in the distal tubule wall

mean arterial pressure (MAP) Average blood pressure in the arteries, estimated as diastolic pressure plus one-third of the pulse pressure

mechanically-gated channel A channel that opens in response to mechanical stimuli such as pressure and heat

mechanistic approach The ability to explain the mechanisms that underlie physiological events

mechanoreceptor A sensory receptor that responds to mechanical energy such as touch or pressure

mediated transport Movement across a membrane with the aid of a protein transporter

medulla Oblongata breathing, cardiovascular control, swallowing, and other unconscious or involuntary functions

membrane potential difference The electrical potential created by living cells due to uneven distribution of ions between the intracellular and extracellular fluids

membrane-spanning protein(s) (also called integral or intrinsic proteins) Membrane proteins that are tightly bound into the phospholipid bilayer

meninges Three layers of membrane that lie between the spinal cord and vertebrae, or brain and skull

menstrual cycle The cyclic production of eggs and cyclic preparation of the uterus for pregnancy in females

mesangial cell(s) Contractile cells in the basal lamina of the renal corpuscle

messenger RNA (mRNA) RNA produced in the nucleus from a DNA template; travels to the cytoplasm to direct the synthesis of new proteins

metabolism All the chemical reactions in the body

metarteriole(s) Branches of arterioles that regulate blood flow through the capillaries

micelle Small droplet of phospholipid, arranged so that the interior is filled with hydrophobic fatty acid tails

microcirculation The arterioles, capillaries and venules

microvilli Finger-like extensions of the cell membrane that increase the surface area for absorption of material

mitochondria The organelles where most ATP is generated

mitosis Cell division that results in two identical diploid daughter cells

molecule Two or more atoms linked together by sharing electrons

motor end plate The specialized postsynaptic region of a muscle fiber

motor unit Group of skeletal muscle fibers and the somatic motor neuron that controls them

multiunit smooth muscle Smooth muscle in which cells are not linked electrically and each muscle fiber is controlled individually

muscle tone The basal state of muscle contraction or tension that results from tonic activity

myofibril(s) Bundles of contractile and elastic proteins responsible for muscle contraction

myogenic contraction Contraction that orig-inates within the muscle fiber as a result of stretch

myoglobin Oxygen-binding pigment in muscle that transfers oxygen between cell membrane and mitochondria

myosin Forms thick filaments of the myofibril that convert chemical bond energy of ATP into motion

myosin light chain kinase (MLCK) Enzyme that phosphorylates light protein chains of myosin in smooth muscle

myosin light chain phosphatase Enzyme that dephosphorylates light protein chains of myosin in smooth muscle

negative feedback A homeostatic feedback loop designed to keep the system at or near a setpoint

nephron Microscopic tubule that is the functional unit of the kidney

nerve A collection of axons running between the central nervous system and the peripheral target cells

neurohormone A hormone that is produced and secreted by a neuron

neuromuscular junction The synapse of a somatic motor neuron and a skeletal muscle fiber

neuron A nerve cell, capable of generating and transmitting electrical signals

neurotransmitter A chemical signal released by a neuron that influences the neuron's target cell

nicotinic receptor A type of acetylcholine receptor that also responds to nicotine

nitric oxide (NO) A short-acting paracrine that relaxes smooth muscle

nociceptor A sensory receptor associated with pain

nodes of Ranvier Unmyelinated regions on myelinated axons

nonpolar molecule A molecule whose electrons are distributed so evenly that there are no regions of partial positive or negative charge

norepinephrine Primary neurotransmitter of the sympathetic division of the nervous system

nucleotide A biomolecule used for storage and transfer of information and energy; composed of phosphate groups, a five-carbon sugar, and a base

nucleus (1) The center of the atom, composed of protons and neutrons; (2) Central body of a cell that contains DNA; (3) A cluster of nerve cell bodies in the central nervous system

organ Group of tissues that carries out certain functions

organelle(s) Assorted intracellular structures that each take on one or more of the cell's functions

organic compound(s) Molecules that contain carbon

origin of a muscle The end of the muscle attached closest to the trunk or to the more stationary bone

osmolality Concentration expressed in osmoles per kilogram

osmolarity Concentration expressed in osmoles per liter

osmosis The movement of water down its concentration gradient across a membrane

oxidation-reduction reaction Involves the transfer of electrons or protons (H+) between chemicals

oxidative phosphorylation Mitochondrial path-way that consumes oxygen and high-energy electrons and produces ATP and water

oxygen consumption The disappearance of oxygen during oxidative phosphorylation, when oxygen combines with hydrogen in the mitochondria to form water

oxytocin Posterior pituitary hormone that causes uterine and breast smooth muscle contraction

paracrine A chemical secreted by a cell that acts on cells in the immediate vicinity

parasympathetic branch Division of the autonomic nervous system that is responsible for day-to-day activities

parathyroid hormone (parathormone, or PTH) Hormone from the parathyroid glands that increases plasma Ca^{2+} concentration

parietal cells Cells of the stomach that secrete hydrochloric acid

partial pressure The pressure of a single gas species

passive transport Movement across a membrane that does not depend on an outside source of energy

peptidase Enzyme that breaks up peptides into smaller peptides or amino acids

pericardium A tough membranous sac that encloses the heart

pericytes Cells that form a meshlike outer layer between the capillary endothelium and the interstitial fluid

peripheral nervous system All neurons that lie completely or partially outside the central nervous system

peripheral receptor(s) Sensory receptors that are not located in or close to the brain

peripheral resistance Resistance to blood flow created primarily by the arterioles

peristalsis Waves of contraction that move along the gastrointestinal tract

peritoneum A membrane that lines the abdomen

permissiveness One hormone cannot exert its effects fully unless a second hormone is present

pH A measure of the concentration of H+; pH = $-$log [H+]

phagocytosis The process by which a cell engulfs a particle into a vesicle by using the cytoskeleton to push the membrane around the particle

pharmacomechanical coupling Contraction that occurs in smooth muscle as a result of a ligand binding; not accompanied by a change in membrane potential

phosphocreatine Muscle molecule that stores energy in high-energy phosphate bonds

phospholipase C Enzyme that converts a membrane phospholipid into two different second messenger molecules

phosphorylation Addition of a phosphate group to a molecule

physiology The study of the normal functioning of a living organism and its component parts

plasma The fluid portion of the blood

plasma membrane The cell membrane that serves as both a gateway and a barrier for substances moving into and out of the cell

platelet(s) Cell fragments that participate in coagulation

podocyte(s) Specialized epithelial cells in Bowman's capsule that surround each capillary

polar molecule(s) Molecules that develop regions of partial positive and negative charge when one or more atoms in the molecule have a strong attraction for electrons

portal system A specialized region of the circulation consisting of two capillary beds directly connected by a set of blood vessels

positive feedback loop A feedback loop in which the response reinforces the stimulus, triggering a vicious cycle of ever-increasing response

postabsorptive state (fasted state) A catabolic state, in which the body taps into its stored reserves and the cells degrade large molecules into smaller molecules

posterior pituitary gland An extension of the brain that secretes neurosecretory hormones made in the hypothalamus

postganglionic neuron Autonomic neuron that has its cell body in the ganglion and sends its axon to the target tissue

potential energy Stored energy that has the ability to do work

power stroke Movement of the myosin head that is the basis for muscle contraction

preganglionic neuron Autonomic neuron that originates in the central nervous system and terminates in an autonomic ganglion

preload The degree of myocardial stretch created by venous return

preprohormone Inactive molecule composed of one or more copies of a peptide hormone, a signal sequence, and other peptide sequences that may or may not have biological activity

presynaptic cell The cell releasing neurotransmitter into a chemical synapse

primary active transport Movement across a membrane that requires the input of energy from ATP

prohormone Inactive hormone which must be chemically modified to become active

proprioception Awareness of body position in space and of the relative location of body parts to each other

prostaglandin(s) Lipid-derived molecules that act as physiological regulators

protease(s) Enzymes that break proteins up into smaller peptides

protein kinase(s) Enzymes that transfer a phosphate group from ATP to a protein

pulmonary circulation That portion of the circulation that carries blood to and from the lungs

pulmonary trunk The single artery that receives blood from the right ventricle; splits into left and right pulmonary arteries

pulmonary valve The semilunar valve between the right ventricle and the pulmonary trunk

pyloric sphincter (pylorus) The sphincter separating the stomach and small intestine

reabsorption Movement of filtered material from the lumen of the nephron to the blood

receptor (1) A cellular protein that binds to a ligand; (2) A cell or group of cells that continually monitor changes in the internal or external environment

receptor potential The graded potential that develops in a sensory receptor

recruitment Addition of motor units to increase the force of contraction in a muscle

relative refractory period A period of time following an action potential during which a higher-than-normal stimulus is required to start another action potential

renin Peptide secreted by juxtaglomerular cells that converts angiotensinogen into angio-tensin

respiration (1) Cellular use of oxygen and substrates to produce energy; (2) Exchange of gases between the atmosphere and the cells

respiratory cycle An inspiration followed by an expiration

sarcomere The contractile unit of a myofibril

sarcoplasmic reticulum Modified endoplasmic reticulum in muscle that concentrates and stores Ca^{2+}

saturation All active sites on a given amount of enzyme or transporter are filled with substrate and reaction rate is maximal

Schwann cell(s) Cells that form myelin around axons of peripheral neurons

second messenger(s) Intracellular molecules that translate the signal from a first messenger into an intracellular response

secondary (indirect) active transport The energy for transport is the potential energy stored in a concentration gradient; indirectly depends on energy of ATP

secondary sex characteristic(s) Features of the body, such as body shape, that distinguish males from females

secretin Intestinal hormone that stimulates bicarbonate secretion and pepsin release; inhibits gastric acid

secretion (1) The movement of selected molecules from the blood into the nephron; (2) The process by which a cell releases a substance into the extracellular space

sensory neuron A neuron that transmits sensory information to the central nervous system

signal transduction The transmission of information from one side of a memebrane to the other using membrane proteins

single-unit smooth muscle Smooth muscle fibers that are electrically coupled by numerous gap junctions

sinoatrial node (SA node) A group of autorhythmic cells in the right atrium of the heart; the main pacemaker of the heart

skeletal muscle Striated muscle usually attached to bones; responsible for positioning and movement of the skeleton

slow wave potential Cyclic depolarization and repolarization of membrane potential in smooth muscle

smooth muscle Muscle found in internal organs that lacks organized sarcomeres

sodium-potassium ATPase (Na+/K+ -ATPase) Active transporter that moves Na+ out of the cell and K+ into the cell against their respective concentration gradients

solubility The ease with which a molecule or gas dissolves in a solution: The more easily a substance dissolves, the higher its solubility

somatic motor neuron(s) Efferent neurons that control skeletal muscles

somatomedin(s) Insulin-like growth factors synthesized by the liver

specificity The ability of an enzyme or receptor to bind to a particular molecule or a group of closely related molecules of control system

spinal reflex A simple reflex that can be integrated within the spinal cord without input from the brain

Starling's law of the heart Within physiological limits, the heart pumps all the blood that returns to it

steroid(s) Lipid-related molecules derived from cholesterol

stimulus The disturbance or change that sets a reflex in motion

striated muscle(s) Muscles that appear to have alternating light and dark bands running perpendicular to the length of the fiber; includes skeletal and cardiac muscle

stroke volume The amount of blood pumped by one ventricle during one contraction

surface tension of water The hydrogen bonds between water molecules that make it difficult to separate water molecules

surfactant Chemical that decreases the surface tension of water

sympathetic branch Division of the autonomic nervous system that is responsible for fight-or-flight response

symport Movement of two or more molecules in the same direction across a membrane by a single transporter

synapse Region where a neuron meets its target cell

synergism Interaction of two or more hormones or drugs that yields a result that is more than additive

systemic circulation Portion of the circulation that carries blood to and from most tissues of the body

systolic pressure The highest pressures in the arteries in a cardiac cycle

teleological approach Describing physiological processes by their purpose rather than their mechanism

tension The force created by a contracting muscle

testosterone Steroid sex hormone, dominant in males

thick filament An aggregation of myosin in muscle

thin filament An actin-containing filament of the myofibril

threshold (1) The minimum depolarization that will initiate an action potential in the trigger zone; (2) The minimum stimulus required to set a reflex response in motion

thyroglobulin Large protein in which thyroid hormones are formed

thyroid gland Endocrine gland in the neck that produces thyroid hormones

thyroxine Tetraiodothyronine or T_4; main hor-mone produced by the thyroid gland

tidal volume (V_T) The volume of air that moves in a single normal inspiration or expiration

tight junction Cell-to-cell junction in epithelia that does not allow much movement of material between the cells

tissue A collection of cells, usually held together by cell junction, that works together to achieve a common purpose

total pulmonary ventilation The volume of air moved in and out of the lungs each minute

transport maximum (T_m) The maximum transport rate that occurs when all carriers are saturated

transverse tubule(s) (t-tubules) Invaginations of the muscle fiber membrane, associated with the sarcoplasmic reticulum

triiodothyronine (T_3) Most active form of thyroid hormone; produced mostly in peripheral tissues from T_4

trophic hormone Any hormone that controls the secretion of another hormone and stimulates the growth and function of the secreting gland

tropomyosin A regulatory protein that blocks the myosin-binding site on actin

troponin Complex of three proteins associated with tropomyosin

tubuloglomerular feedback The process by which changes in fluid flow through the distal tubule influence glomerular filtration rate

type I alveolar cell(s) Thin alveolar cells for gas exchange

type II alveolar cells(s) Alveolar cells that synthesize and secrete surfactant

tyrosine kinase Membrane enzyme that adds a phosphate group to the tyrosine residue of a cytoplasmic protein, enhancing or inhibiting its activity

van der Waals force Weak attractive force that occurs between two polar molecules or a polar molecule and an ion

varicosity Swollen regions at the distal ends of autonomic axons that store and release neurotransmitter

vasa recta Peritubular capillaries in the kidney that dip into the medulla and then go back up to the cortex, forming hairpin loops

vascular smooth muscle The smooth muscle of blood vessels

vasoconstriction Contraction of circular vascular smooth muscle that narrows the lumen of a blood vessel

vasodilation Relaxation of circular vascular smooth muscle that widens the lumen of a blood vessel

velocity of flow The distance a fixed volume of blood will travel in a given period of time

venous return The amount of blood that enters the heart from the venous circulation

ventilation The movement of air between the atmosphere and the lungs

ventral root Section of a spinal nerve that carries information from the central nervous system to the muscles and glands

ventricle Lower chamber of the heart that pumps blood into the blood vessels

vesicle A sac-like, membrane-bounded organelle used for storage and transport

vesicular transport The movement of vesicles within the cell with the aid of the cytoskeleton

viscosity Thickness or resistance to flow of a solution

vital capacity The maximum amount of air that can be voluntarily moved in and out of the respiratory system

voltage-gate channel A gated channnel that opens or closes in reponse to a change in membrane potential